THE PEDAGOGIES OF RE-USE

The Pedagogies of Re-Use captures the amazing digital gathering of students, academics, practitioners, and activists that happened at the International School of Re-Construction. Involving over 100 people, from countries as far apart as Brazil, Canada, Ireland, UK, Spain, Germany, Greece, UAE, and China, the participants spent two weeks working in eleven teams to consider architectural propositions responding to the current climate and ecological emergency. This book documents the work of the eleven teams, considering the themes they pursued, the student projects proposed, and the final design ideas developed by each group. Supplemented with images of the work, the book also includes leading academics and professionals who supported the school and contribute their voices to these crucial issues of deconstruction, re-use, and adaptation. It is ideal reading for students and academics looking at the issues created by the climate emergency to which architecture must respond.

Duncan Baker-Brown is a practicing architect, academic, and environmental activist. Author of *The Re-Use Atlas: A Designer's Guide Towards a Circular Economy*, he has practised, researched, and taught around issues of sustainable development and closed-looped systems for more than 25 years. He recently founded BakerBrown, a research-led architectural practice and consultancy created to address the huge demands presented by the climate and ecological emergency as well as the challenges of designing in a post-COVID world. Over the years Duncan's practices (and academic 'live' projects) have won numerous accolades including RIBA National Awards and a special award from The Stephen Lawrence Prize for the Brighton Waste House – the prize money has since been used to set up a student prize for circular, closed-loop design at the University of Brighton, UK, where Duncan teaches.

Duncan was the University of Brighton's principal investigator for the North West Europe's Interreg Facilitating the Circulation of Reclaimed Building Elements (FCRBE) project. He was responsible for curating the pedagogic outputs for the FCRBE team (lead by Rotor). Said outputs are the subject of this book, which he has co-edited with the wonderful Prof. Graeme Brooker.

Graeme Brooker is Professor and Head of Interiors at The Royal College of Art, London, UK. He has published numerous books on many aspects of the interior including the recent publications *50|50 Words for Reuse* (2022), *Brinkworth: So Good So Far* (2019), *Adaptations* (2016), and *Key Interiors Since 1900* (2013). He has co-authored/edited ten books on the interior including the highly acclaimed *Rereadings* (2005; Volume 2, 2018). He has led interior programmes in Cardiff, Manchester, Brighton, and London institutions and has been a visiting professor in Antwerp, Berlin, Istanbul, and Milan. He is a member of the editorial advisory board of the journals *Interiors: Design: Architecture: Culture*, *INNER*, *IDEA,* and *DESIGN&*. He is the founder and was the director of the charity Interior Educators (IE), the national subject association for all interior courses in the UK, between 2006–2018 and 2023–present. He is a trustee of United In Design (UID), a charity set up to address the lack of diversity in the profession of interiors. He is currently working on the funded project – ATLAS, an archival-based work with the European Council of Interior Architects (ECIA) and the books *The SuperReuse Manifesto* (2024) and *The Story of the Interior* (2025). The latter is a history book that moves beyond standard chronological accounts and, instead, retells thematic histories of inside spaces through narratives of the room and the private and public interior.

THE PEDAGOGIES OF RE-USE

The International School of Re-Construction

Edited by Duncan Baker-Brown and Graeme Brooker

Designed cover image: Tim Danson, Architectural Technology student, School of Architecture Technology & Engineering, University of Brighton, UK.

First published 2024
by Routledge
4 Park Square, Milton Park, Abingdon, Oxon OX14 4RN

and by Routledge
605 Third Avenue, New York, NY 10158

Routledge is an imprint of the Taylor & Francis Group, an informa business

© 2024 selection and editorial matter, Duncan Baker-Brown and Graeme Brooker; individual chapters, the contributors

The right of Duncan Baker-Brown and Graeme Brooker to be identified as the authors of the editorial material, and of the authors for their individual chapters, has been asserted in accordance with sections 77 and 78 of the Copyright, Designs and Patents Act 1988.

The Open Access version of this book, available at www.taylorfrancis.com, has been made available under a Creative Commons Attribution-Non Commercial-No Derivatives (CC-BY-NC-ND) 4.0 license.

Trademark notice: Product or corporate names may be trademarks or registered trademarks, and are used only for identification and explanation without intent to infringe.

British Library Cataloguing-in-Publication Data
A catalogue record for this book is available from the British Library

Library of Congress Cataloging-in-Publication Data
Names: Baker-Brown, Duncan, editor. | Brooker, Graeme, editor. |
School of Re-Construction (2021 : Online)
Title: The pedagogies of re-use : the International School of Re-construction /
edited by Duncan Baker-Brown and Graeme Brooker.
Description: Abingdon, Oxon : Routledge, 2024. | Includes bibliographical references and index.
Identifiers: LCCN 2024001653 (print) | LCCN 2024001654 (ebook) |
ISBN 9781032665511 (hardback) | ISBN 9781032650623 (paperback) |
ISBN 9781032665559 (ebook)
Subjects: LCSH: Architecture and climate. | Sustainable architecture. |
Buildings—Remodeling for other use.
Classification: LCC NA2541 .P43 2024 (print) | LCC NA2541 (ebook) |
DDC 720.47—dc23/eng/20240508
LC record available at https://lccn.loc.gov/2024001653
LC ebook record available at https://lccn.loc.gov/2024001654

ISBN: 978-1-032-66551-1 (hbk)
ISBN: 978-1-032-65062-3 (pbk)
ISBN: 978-1-032-66555-9 (ebk)

DOI: 10.4324/9781032665559

Typeset in Times New Roman
by Apex CoVantage, LLC

The Pedagogies of Re-Use is part of an European Union European Regional Development Fund (EU ERDF) £4.33 million Interreg North West Europe (NWE) project titled 'Facilitating the Circulation of Reclaimed Building Elements' (FCRBE), Interreg NWE 739, October 2018–December 2023. Online publication: June 2024, London.

The FCRBE project aims to increase the amount of reclaimed building elements in circulation within its territory by +50% (in mass) by 2032.

Useful Links

FCRBE: https://vb.nweurope.eu/projects/project-search/fcrbe-facilitating-the-circulation-of-reclaimed-building-elements-in-northwestern-europe/

Opalis: Building and renovating with reclaimed materials, https://opalis.eu/en

Facilitating the Circulation of Reclaimed Building Elements in Northwestern Europe – The University of Brighton: https://research.brighton.ac.uk/en/projects/facilitating-the-circulation-of-reclaimed-building-elements-in-no

School of Re-Construction: https://blogs.brighton.ac.uk/schoolofreconstruction/

This publication reflects the authors' views only. The authors and the funding authorities are not liable for any use that may be made of the information contained therein.

This document benefited from the financial support of the European Regional Development Fund through the Interreg NWE programme.

bruxelles
environnement
leefmilieu
brussel
.brussels

Buildwise

SALVO
SalvoWEB.com

City of Utrecht

CSTB
le futur en construction

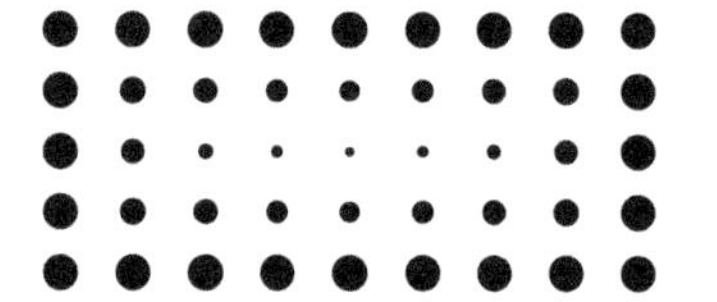

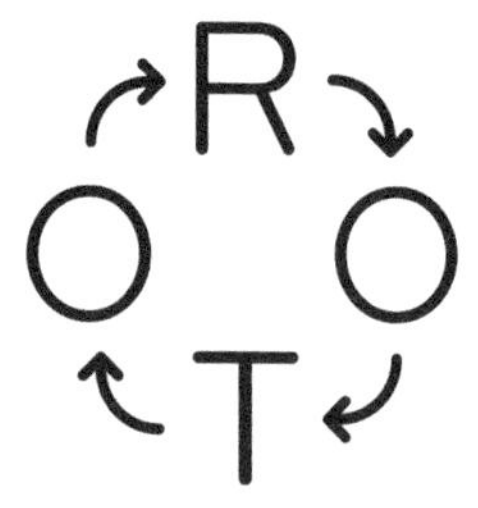
R
O O
T

Embuild

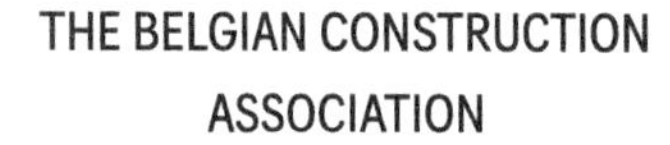
THE BELGIAN CONSTRUCTION
ASSOCIATION

LUXEMBOURG
INSTITUTE OF SCIENCE
AND TECHNOLOGY
LIST

TU Delft
Delft University of Technology

EIS

University of Brighton

CONTENTS

CONTRIBUTORS

Duncan Baker-Brown was the University of Brighton's principal investigator for the North West Europe's Interreg1 Facilitating the Circulation of Reclaimed Building Elements (FCRBE) project. He was responsible for curating the pedagogic outputs for the FCRBE team (lead by Rotor). Said outputs are the subject of this book, which he has co-edited with the wonderful Prof. Graeme Brooker.

Photography Laura Macaulay, 2019

Eddie Blake is an architect and academic. He trained at the Mackintosh School of Architecture and the Architectural Association. He teaches architecture at the London School of Architecture. He previously taught at the Royal College of Art, and the Bartlett (UCL). Blake is a practicing architect and Director at Studio Weave. He previously has worked as an architect and designer at renowned international practices: Sauerbruch Hutton and Sam Jacob Studio. He writes about architecture and culture for the Guardian and Real Review among other publications.

Photography Piers Alladyce, 2020, © RCA

Graeme Brooker is Professor and Head of Interiors at The Royal College of Art, London, UK. He has published numerous books on many aspects of the interior including the recent publications *50|50 Words for Reuse* (2022), *Brinkworth: So Good So Far* (2019) *Adaptations* (2016), *Key Interiors Since 1900* (2013), and the highly acclaimed *Rereadings*, (2005, Volume 2 2018).

Photography Raphael Thibodeau, 2016

Louis Destombes is a French-graduated architect and holds a PhD in architecture. His professional practice within the coop Bellastock embraces applied research, expertise, and project coordination in the fields of circular economy and transitory urbanism. An associate teacher at Ecole nationale supérieure d'Architecture Paris-La Villette and researcher at AHTTEP, his academic works and teaching focusses on the history of construction theory in architecture and the mutations of architects' constructive culture along with the environmental turn of building practices.

Photography, Lukas Turcksin, 2021. Copyright granted by Photo Lukas Turcksin

Lionel Devlieger is an engineer-architect and historian. In 2006, he co-founded Rotor with Maarten Gielen and Tristan Boniver. Rotor is a Brussels-based organisation specialised in the study of present-day material culture. Lionel has been teaching in universities across Europe and the US (UC Berkeley, TU Delft, Columbia University, the AA School). He co-authored Deconstruction et reemploi, a reference textbook on building component reuse (EPFL press, 2018) and Ad Hoc Baroque, on Marcel Raymaekers, a post-war pioneer in design with architectural salvage (Rotor, 2023). Since 2021, Lionel is fulltime associate professor in the material-cultural history of architectural practice at Ghent University

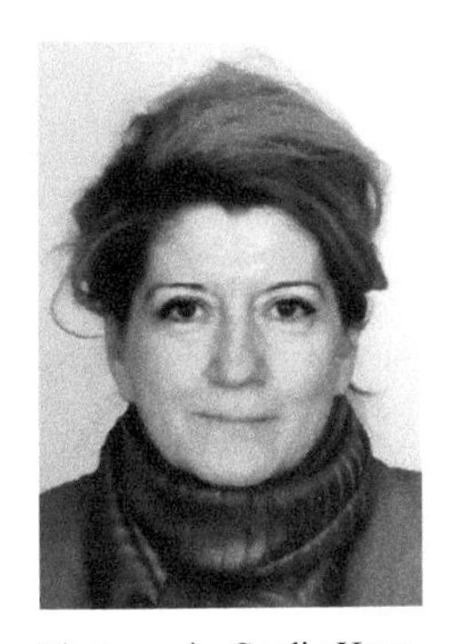

Photography Studio Heno, 2020 Copyright

Elma Durmišević is a visiting professor at the Polytechnic University of Milan, Italy. She is a leading architecture authority on reversible/circular building design and transformable buildings in Europe. Elma holds a PhD at Delft University of Technology (TU-Delft) on transformable building structures and design for disassembly in architecture. As an associate professor at TU–Twente, Durmišević introduced master program 'green engineering in architecture' through design studios.

Durmišević is founder of 4D Architects Amsterdam, founding director of EU Laboratory for Green Transformable Buildings (GTB Lab) in the Netherlands, and creator of Green Design Centre for Southeast Europe in Mostar, Netherlands.

During more than 20 years of innovation and leading circular building developments within various EU projects, Durmišević developed Circular Building Design Toolkit including design guidelines, protocol, and tools for reversible buildings, which are adopted as EU guidelines and form the circular building design checklist of the Brussels Region. These tools measure circularity of buildings based on disassembly, re-use potential, and transformation capacity, resulting into Circular Building Profiles and Circular DigitalTwin ReversibleBIM.

Photography Claire Hoskin

Nick Gant is an award-winning sustainable designer and published researcher whose consultancy and research impacts global, industrial companies; international NGOs; and governing bodies and civic communities in the UK and overseas.

He has collaborated with or designed for international industrial brands and NGOs including PUMA, New Balance, The BBC, Vivienne Westwood, Interface, Tesco, The Beatles and Apple Corps, Vision Aid Overseas, The Marine Conservation Society, The NHS, The British Council, Philips, The V&A and Natural History Museum, Ellen MacArthur Foundation, and The Design Council.

He disseminates research internationally on subjects including design for circular economies; the prevention of waste; material innovation; health and well-being; and sustainable development across areas of fashion, product, and industrial design and the built environment.

Photography Victor Pattyn, 2023

Maarten Gielen is co-founded Rotor in 2006 where he worked as a designer, researcher, and manager. Drawing on a rich experience in deconstruction, logistics, architecture, and building design with salvaged materials, Maarten also co-founded and steered the development of the spin-off Rotor DC (Deconstruction), where he occupied the position of director until 2023. Maarten played a key role in most exhibitions and publications produced by Rotor such as Deutschland Im Herbst (2008), Usus/Usures for the Belgian pavilion at the 2010 Venice Biennale, Ex-Limbo for the Fondazione Prada in Milan (2011), OMA/Progress, on the work of Office for Metropolitan Architecture (OMA)/Rem Koolhaas at the Barbican Art Gallery in London, and Behind the Green Door for the 2013 Oslo Architecture Triennale. For his contribution as an author, designer and projects initiator. Maarten received the Maaskant Prize for Young Architects in 2015. He has taught in many architecture and design schools including HEAD in Geneva, the Technical University (TU) in Delft, the Karlsruhe Institute of Technology, and Columbia University in New York. Maarten recently founded Halfwerk.be, a manufacturer of objects in reclaimed steel.

Photography Claire Hoskins, 2019

Lucy-Ann Gilbert is trans-disciplinary designer, maker, educator, and allotmenteer. Through her work she strives to explore the boundaries of spatial design and has worked in a range of settings including film, exhibition design, collaborative design through making, architectural practice alongside working in education. From 2016 to 2023, she taught and later led the Interior Architecture course at the University of Brighton.

By taking a trans-disciplinary approach she invites potentially new and unexpected connections to be made through the design process – something she feels is essential in order to respond to the ever-changing needs and challenges of our lived and built environments.

Photography Rotor, 2021

Michaël Ghyoot and Sophie Boone. Founded in 2005, Rotor is a non-profit association that investigates the organisation of the material environment. Rotor develops critical positions through research and design. Besides projects in architecture and interior design, Rotor also produces exhibitions, books, economic models and policy proposals. The workshop was animated by Sophie Boone and Michaël Ghyoot. Both are project managers at Rotor and active in a wide range of activities: research, design, assistance to building owners, and exhibitions making.

Photography Michael Howe, 2019

Michael Howe comes to academia from architectural practice, where he has over 28 years' experience working for highly regarded architectural practices such as Zaha Hadid Architects, Patel Taylor Architects, and Matthew Lloyd Architects. In 2001, he co-founded Mae Architects with Alex Ely, as a housing design specialist practice. Projects ranged in size from small arts and domestic projects (under 100 msq), to urban-scale regeneration and housing design. Other activities included writing for the architectural press, client training, policy development, exhibitions, and teaching. Originally trained as a fine artist at Central School of Art, London. Michael has been a fully qualified architect for 21 years.

Photography Eric Toribio, 2020

Taleen Josefsson is an American-registered architect and passive house consultant who applies a holistic approach to tackling the built environment's role in the twin climate and biodiversity crises. She is particularly focussed on the specification of resources and materials in construction utilising circular economy principles, including assessing materials' life cycles, chemical content, embodied carbon, and re-use potential. Taleen works as a Thrive Project Leader at Chetwoods, advising on sustainable and circular practice and implementation both within the firm and for external clients. Outside of the office, she is a coordinator of the ACAN (Architects Climate Action Network) Circular Economy group.

Photography Kurt Mundahl, 2019

Folke Köbberling is professor and head of the Institute for Architecture-Related Art at the Technical University of Braunschweig, Germany. In her artistic work, she formulates forms of resistance against our usurpation by the excesses of the ruling neoliberal economic order, such as the compulsion to consume and the destruction of cities, landscapes, public spaces, resources, and people. In addition to an engagement with pure market interests and the radical urban transformation which Berlin has undergone since reunification, her focus centres on automobile individual transport as a hegemonic guiding culture, which she comments on using artistic means as well as ironically subverting and pointing out its limits and alternatives. With Martin Kaltwasser, she has built the Jellyfish Theatre in 2011 in London, which was nominated for the Mies van der Rohe Award and got the First Prize in *Architects' Journal* Small Buildings Sustainability Award.

Photography Kurt Mundahl, 2019

Alexa Kreissl studied sculpture in Paris, France, and Düsseldorf, Germany. As an architecture-related artist, she investigates urban space, its dynamics in constant transition, its niches, wastelands, opportunities, and resources. Her modular adaptive structures are suggestions for use that respond to situational circumstances and needs, which imply transformation and re-purposing. She has made research trips and residencies in Paris, Japan, California, and New York. An artist and PhD candidate, she is a researcher at the Institute of Architecture-Related Art, Department of Architecture of the Technical University Braunschweig (Germany) and a research associate at the Leibniz Science Campus – Postdigital Participation – Braunschweig (LSC PDP) in the project 'People, Digital Intelligence and Recycling – Designing Urban Life together'. Refuse resources, art, architecture, design, and aesthetic transformation are the starting point for her interdisciplinary and experimental work.

Photography Perry Schrijvers, Yr, © Maurer United Architects

Nicole Maurer is educated as architect and urbanist and has been active as creative director of Maurer United Architects since 1998. In addition, she teaches students in Europe and China and plays an active part in several cultural and social organisations.

James McAdam is a designer with a wide knowledge of the 'digital toolbox' available to designers, makers, and architects, including 2D CAD, BIM, 3D modelling, animation, CAM, and presentation. James studied MA in design products at the Royal College of Art, UK, in 2002 and has since designed commercially for exhibitions, retail VM (Visual Merchandising), architectural scale art installations, publishing, and pyrotechnic products. Over the last 17 years, he has mixed teaching, offering 'real world' experience to degree students. He is currently focusing on his 'DigiSkills' teaching within architecture at the University of Brighton and Design/Craft BA. He has just been granted a worldwide patent on one of his designs for a pyrotechnic ignition system and enjoys spending time at the workbench as much as at the keyboard and mouse.

Photography Copeland, 2017

Scott McAulay. With roots in Scottish climate justice activism, Scott's practice through the Anthropocene Architecture School (AAS) fuses architectural, and political education with Climate Literacy, Spatial Justice, and the Radical Imagination. Through temporary activations of space – from festivals to peoples' assemblies, their writing, teaching, and educational workshops explore where architecture, construction, and housing fit into a compassionate Just Transition. Their work through the AAS – an agile infrastructure for un/learning, and otherwise, prefiguratively challenges the architectural education system and the wider construction industry's ongoing inertia in the face of today's cascading climate emergency and cost of survival crisis.

One of 2020's RIBAJ Rising Stars, Scott has taught widely across Scotland, the British Isles, mainland Europe, and Turtle Island. Scott was a founding member of ACAN Scotland, a past coordinator of ACAN's Climate Literacy Working Group, an active member of Living Rent – Scotland's tenants union, and has worked architecturally with Architype since 2021 – deploying specialisms in ecological materials and low carbon design across projects at home in Scotland and across the wider UK, cultivating a fluency in lifecycle analysis and leading on the development of the studio's Climate Action Roadmap. Having launched the AAS on its Open Programme in 2019, Scott joined the Architecture Fringe team as a Co-Producer in 2021 and in 2023, he was part of co-creating the Retrofit Reimagined festival season.

Photography Liz Gorman, © 2019

Inês Neto dos Santos is a multidisciplinary artist, working with food, people, and spaces as metaphors for symbiotic relationships to our surroundings. Her practice moves between performance, installation, and social sculpture, investigating the socio-political implications of what we eat and how we come to eat it – creating contexts through which to explore and discuss ecology, collaboration, and togetherness. Recently, her research has taken her into the study of beans and their soil regeneration qualities, investigating their relationship to supported sustenance and care. Inês teaches, writes, and cooks, having been a guest lecturer at Kingston University and Westminster University in the UK. Since March 2021, alongside artist Nora Silva, she co-leads the online course 'Food Cosmogonies'.

Photography Job Boersma, 2023

Mark Oldengarm is educated in history and history of arts and worked as program manager of sustainability at the Royal Institute of Dutch Architects. Currently he is the business manager for two thrift shop organisations, owns a social and sustainable advising business, and is a council member in the city of Zwolle, Netherlands.

Photography Peter Cathersides, 2021

Filipa Oliveira is a London-based architect who advocates for circularity within the industry and is passionate about mitigating the impacts of construction on the natural environment. She promotes regenerative design and a holistic and multidisciplinary approach to tackle the complex environmental and social issues that we are currently facing.

She graduated with a master's degree in architecture and urban planning in Portugal, where she developed an avid interest in research, circular cities, and activism. Her interests lie in the re-use of materials, vernacular architecture, and urban regeneration within its culture, time, and space.

Filipa is a steering group member for the Architects Climate Action Network (ACAN) in the UK and Portugal, where she coordinates and helps research on circular economy.

Photography Jonny Pugh, 2021

Jonny Pugh is an architect, designer, and educator working between Porto (Portugal), Barcelona (Spain) and London (UK). His work is focussed on participatory design processes, rehabilitation, and re-use. He has taught/lectured at universities in the UK, Portugal, and Switzerland.

In practice, he has established relationships with international cross-disciplinary design collectives. He is a collaborator with Barcelona-based studio Flores & Prats since 2005, working closely on the themes of re-use and social and material heritage, the focus of presentations at the Venice Architecture Biennale (2018, 2023) through projects such as Sala Beckett. In 2023, he co-founded the Portuguese not-for-profit association NADA NOVO/NO NEWS, dedicated to exploring the cultural challenges of material re-use in construction.

Photography Claire Hoskins, 2019

Anthony Roberts is an architect and academic who describes himself as a single person practice. His architectural/building work is small/mid-scale, quiet in material choice, and intimate in detailing. He also likes to produce 'event-surprises'; where-in material and formal 'twists and turns' to punctuate standard spatial configurations. His architectural life has been 'interrupted' by extended periods of work in other fields; film, painting, graphics, design, photography, and making exhibitions. He does not this work as 'outside' or 'other than' architecture – it's more like 'an-other' architecture.

He has worked with major artists, including Christo, Jan Jelle Stroosma, and David Fox. His exhibitions include 'Talking to the Brick Wall' (1986) and 'Shadow Dancers' (1990) on homelessness and the Gulf War (1993). Since 1998, he has been teaching running parallel with small architecture as a studio tutor of both undergraduate and master's programmes at the School of Architecture and Design (University of Brighton). He is a peripatetic, non-specialist teacher in design, technology, and humanities. He has a great sense of humour and is a wonderful storyteller.

Photography Andrzej Soltysiak, 2019

Katarzyna Sołtysiak. Having graduated from the University of Brighton, UK, and TU Delft, Netherlands, Katarzyna enjoys working on the fringes of architecture. Currently, she is an active architect and part of Team SUM researching circular transformation of social housing in the Netherlands emphasising its impact on communities. Her initial acquaintance with material re-use took place on the building site of the Waste House. Later she pursued the topic alongside Rotor (BE), ArchitectuurMaken (NL) and One Architecture (NL/US). Her interest in materials has led to founding Project Demeter – actively employing social media for education on the topic.

Photography Tanya Southcott, 2023

Ben Sweeting teaches architecture and design at the University of Brighton, UK. Ben's research concerns relationships between cybernetics, systemic design, and architectural theory. After studying architecture at the University of Cambridge and University College London, Ben completed a PhD at the latter, supervised by Neil Spiller and Ranulph Glanville. Ben is one of the coordinators of the Radical Methodologies Research Group, which is concerned with questioning the foundations of research and practice in and beyond design.

Photography Bellastock, 2019

Hugo Topalov is an architect and engineer working within Bellastock, a cooperative organization developing experimental architecture since 2006. Hugo is working on the topic of circular economy in architecture and, more specifically, deconstruction and re-use of building materials. His activities are divided between assisting architects and other professional actors of the construction sectors, participating in research projects, training, and raising awareness among various professional audiences.

Hugo has been an assistant teacher at the Belleville Paris School of Architecture, France, since 2020.

Photography Neil Gonzalez, 2019, @neil.gonzalez

Sam Turner is an architect, educator, and activist. He has worked for several award-winning practices focussing on cross-disciplinary working, technical excellence, and care through design.

With Scott McAulay together they joined forces to form Anthropocene Projects C.I.C., a vehicle to promote understanding of the climate crisis and ecological emergency to those working within the built environment, to highlight their personal and professional impact, and facilitate behavioural changes that will help prevent environmental degradation.

Photography Lilo Bohn, 2018

Andre Viljoen is Professor of Architecture at the School of Architecture, Technology and Engineering at the University of Brighton, UK.

Andre is an architect and academic who since the 1990s has been researching the role that urban agriculture and, more generally, urban food systems can play in shaping cities to be more sustainable and resilient. With Katrin Bohn, he developed the concept of Continuous Productive Urban Landscapes (CPULs), which advocates the coherent introduction of urban agriculture into cities. Significant publications include Bohn and Viljoen's book *CPULs Continuous Productive Urban Landscapes: Designing Urban Agriculture for Sustainable Cities* (2005) and its sequel, *Second Nature Urban Agriculture: Designing Productive Cities* (2014). He has been chair of the Association of European Schools of Planning (AESOP) Food Planning Group and, in 2015, Bohn and Viljoen's CPUL work won the RIBA President's Award for Outstanding University Located Research.

Photography Claire Hoskins, 2019

Ryan Woodard has a PhD in waste management and has been undertaking research, consultancy, and teaching in waste since the mid-1990s. He has completed over 50 projects – most of which have been applied – identifying challenges and developing solutions. This has included trialling different collection systems and behaviour change programmes and evaluating impact. He has been working in South Africa for over a decade and supervised PhD/MPhil students on waste in Bangladesh, China, Nigeria, and Ghana. He is course leader for the MSc in environmental assessment and management and the environmental practitioner degree apprenticeship, and he serves on the editorial panel of the Institution of Civil Engineers journal *Waste and Resource Management*.

SCHOOL OF RE-CONSTRUCTION STUDENT LIST, BY GROUP AND TEAM LEADERS

Group A:	Raw 1
Team Leaders:	Nick Gant and Dr Ryan Woodard
Students:	Olivia Harrison
	Julie Van Raemdonck
	Jason Wan
	Yihan Wang
	Leo Sixsmith
	Bruna Borges
Group B:	Useless 1
Team Leaders:	Jonny Pugh and Eddie Blake
Students:	Zhihan Liu
	Nafisah Musa
	Charissa Leung
	Julia Flaszynska
	Thomas Parker
	Hendrik Ringers
	Manon Ijaz
Group C:	Byproduct
Team Leaders:	Michael Howe and James McAdam
Students:	Huiyun Chen
	Dorianne Dupré
	Melahat Gur
	Murjanah Uwais
	Vinciane Gaudissart
	Yacine Abdelghenine
Group D:	Hybrid
Team Leaders:	Anthony Roberts and Katarzyna Sotysiak
Students:	Hsin-Yun Lai
	Aidana Roberts

Amila Strikovic
Weronika Walasz
Ana Pastor
Xiaomeng Ge
Zainab Murtadhawi Qanawati

Group E: Offcut
Team Leaders: Michaël Ghyoot and Sophie Boone
Students: Wen Yuehan
Okgon (YuKun) Huang
Kaylee (Jade) O'Hagan
Konstanca (Connie) Ivanova
Mima Jupp
William Warr
Emily King

Group F: Material Flows
Team Leaders: Mark Oldengarm and Nicole Maurer
Students William Harvey
Kirsten Mcdove
Seyed Mohammad Hossein Rezvani
Marielisa (Maria Elizabeth) Maldonado
Arjun Ramchurn
Sonja Draskovic
Jack Veaney

Group G: Housing
Team Leaders: Filipa Oliveira and Taleen Josefsson
Students: Arthur Ferreira de Araujo
Katiea (Jekaterina) Ancane
Luke Hardman
Callum Purdue
Divya Chand
Razan Atwi
Simon Schaubroeck
Yuerong Wang

Group I: Raw 2
Team Leaders: Scott McAulay and Sam Turner
Students: Johanna Moro
Marvina Sinjari
Pauline Harou
Katrien Devos
Peter Ivan Monos
Daniela Martins
Yuwei Ren

Group J: Useless 2
Team Leaders: Andre Viljoen and Inês Neto dos Santos
Students: Ugne Neveckaite
Soukaïna Lahlou

	Natalia Hryszko
	Gulcan Shyukryuogl
	Natasha Hromanchuk
Group K:	Raw 3
Team Leaders:	Graeme Brooker, Hugo Topalov, and Louis Destombes
Students:	Tess Hillan
	Timothy Danson
	Katarzyna Podhajska
	Alan Wang Chung Cheng
	Mariam Hesham Abuelsaoud
	Gabrielle Kawa
Group L:	Useless 3
Team Leaders:	Folke Köbberling and Alexa Kreissl
Students:	Rebecca Fish
	Elisabeth Ajayi
	Shemol Rahman
	Orla Fahey
	Kai Moritz Tanner

FOREWORD

These are indeed strange and troubling times. We know the world must be remade, but it is well beyond technical fixes – as though underneath all is well, and all will be well. If we are in the midst of a global 'polycrisis', our needs are more profound, akin perhaps to realising an absence of narrative. George Monbiot expressed it like this: 'Despair is the state we fall into when our imagination fails. When we have no story that explains the present and describes the future, hope evaporates'. This volume can't hope to be an economic and social narrative but *The Pedagogies of Re-Use: The International School of Re-Construction* is a marker, generating insight towards a more satisfactory, satisfying narrative in a key sector, fittingly called *construction.* Let us see the world afresh and build differently. It uses a shift from extractive to circulatory, something which surely needs to apply to both the money and materials cycles. But that is for another day.

Yet . . . just as a circular economy is a systemic approach to managing stocks and flows of products, components, and materials, it is itself embedded in other, often more powerful, systems. Society so often expresses itself through its buildings and how it makes and remakes its cities. In today's era of excess, of immense inequality, of asset bubbles, the disorientating pencil-thin residential skyscrapers cluster together like so many visible cash deposit boxes. They say, 'We are not you!' Everywhere there is also, as if by recognition of other possibilities, a distracting veneer of sustainability, while, proximately, social housing is neglected as the tents of the homeless are considered an affront to decency and economic security declines as fast as housing affordability. Nor is it just what might be expected of a financialised rentier economy feeding on post 2008 (– end 2021) easy money in the West: China has reputedly poured more concrete between 2019 and 2022 than the USA did in the whole of the 20th century.[1]

Malinvestment is rife. *Business Insider* identifies around 65 million surplus apartments in China,[2] and other sources claim another 20 million unfinished or stalled apartments (at mid 2023).[3] Roads, bridges, and high-speed rail to places which can never justify the investment also mark a systemic driver: to grow an economy means one or all of three things – consumption, investment, or exports – which, in turn, benefit from ignoring real costs. And the faster the better. Here is Prof. Paul Ekins from University College London: 'Currently, the mainrule of the game is that the polluter does not pay. We don't pay for carbon, we don't pay for the real cost

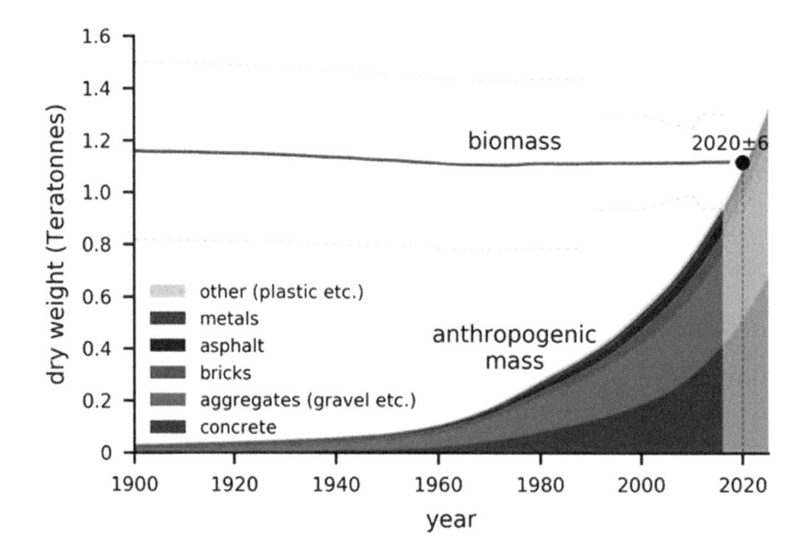

Anthropogenic mass exceeds biomass. By far, the greatest proportion is related to the built environment.

Source: Nature www.nature.com/articles/s41586-020-3010-5 Figure 1 (Dec 2020)

of raw materials, and we don't pay for the environmental impacts'.[4] Ekins was speaking at the World Resources Forum (WRF) in September 2023 where the theme was '*Rethinking Value – Resources for Planetary Wellbeing*'. WRF suggests that making resources a driver for shared well-being within planetary boundaries can be based around:

Sufficiency: from a consumer to a sufficient society
Value chains: from extractive to regenerative value
Digitalisation: from growth to purpose

This is bold and is based upon rethinking and articulating in search of that better narrative. This is a change of mindset. It starts with education, of course. Firstly, by being able to name the problems, even if they are 'wicked' – complex and systemic – and are resolved by a combination of zooming out to grasp the big picture then back in to the detail. This sort of education 'plays true': it holds ideas tightly enough so that a 360-degree perspective is offered but not so tightly that change is excluded or better ideas substituted. After all, education does not assume the ends are given, unlike schooling which does.

The digital School of Re-Construction ran for two weeks in the summer of 2021 as COVID had dislocated plans for the original site-based summer school. Evidenced by the book, a concerted effort was made to bridge across silos and to reflect on the 'mega themes', some of which I touch upon previously.

The Pedagogies of Re-Use: The International School of Re-Construction is embedded in educational enquiry, not schooling, and is creative and critical – discussing failure is a good sign of this – as much as it is informative, drawing on expertise very widely from within the project partners and the 11 teams active under that heading.

Change is coming, and since building and construction – and the circular economy – will be central to any positive outurns at scale, the education stimulus and resource gathered in *The Pedagogies of Re-Use* is as much a labour of love as it is one of constructive optimism.

Access the work of the School of Re-Construction: FCRBE - Facilitating the circulation of reclaimed building elements in Northwestern Europe | Interreg NWE (nweurope.eu) https//vb.nweurope.eu/projects/project-search/fcrbe-facilitating-the-circulation-of-reclaimed-building-elements-in-northwestern-europe/#tab-9

Dr Ken Webster
November 2023
Ken Webster is Visiting Professor at Cranfield University, Fellow of the Cambridge University Institute for Sustainability Leadership (CISL), and a supervisory board member of the digital passport organisation The Madaster Foundation (Netherlands) focussed on the built environment. He was Head of Innovation at the Ellen MacArthur Foundation (2010–2018), a pioneer circular economy NGO.

Notes

1 www.sustainabilitybynumbers.com/p/china-us-cement
2 www.businessinsider.com/china-empty-homes-real-estate-evergrande-housing-market-problem-2021-10?r=US&IR=T
3 www.cnbc.com/2023/11/15/chinas-unfinished-property-projects-are-20-times-the-size-of-country-garden.html
4 *Rethinking Value: Resources for Planetary Wellbeing*, World Resources Forum (Sep 2023) p13. www.wrforum.org

PART I

Introductory essays

INTRODUCTION TO THE SCHOOL OF RE-CONSTRUCTION

Duncan Baker-Brown and Graeme Brooker

In 2012, the United Nations published data stating that humans consumed between 45–60 billion tonnes of mined and harvested raw materials in 2010. Fast forward to the World Economic Forum in Davos in 2021 where *The Circular Gap Report 2020* was presented. It confirmed, among other things, that 2020 was the first time ever that the global economy had consumed over 100 billion tonnes of raw materials, described as "minerals, fossil fuels, metal and biomass." It also noted that the world's economy was only 8.6% circular, down from 9.1% in 2018, the year this annual report was first launched.

So why do these unfathomably large statistics need to concern us all working, as we do, in the world of designing and constructing the built environment? Well, quite simply, the processes associated with extracting and harvesting these raw materials are destroying precious biodiverse landscapes and oceans at an unprecedented rate. The fact that the expansion of these destructive practices has almost doubled in a decade and, post-COVID, has accelerated past the 100 billion tonnes a year mark is especially concerning.

So why am I opening this introduction with the previous statement? The field within which most architects, designers, engineers, and surveyors work, that is, the construction sector, is responsible for consuming approximately 50% of all raw material flows. In the UK, where I am based, our sector also creates 62% of all annual UK waste flows, which amounts to 120 million tonnes of construction-rated material going to landfill and (mostly) incineration. That compares very poorly with the rest of the European Union where construction-related waste flows range from 35–45% of nation totals, but those lower figures are still a big deal, with the reduction in consumption of resources rapidly becoming a high priority across Europe and the UK.

Over the last couple of years, since we emerged out of the COVID pandemic, issues relating to resource consumption have become more apparent. This has been heightened by numerous energy crises around the globe, massive inflation in the cost of materials, and the lack of available labour – all resulting in a cost-of-living crisis. To that, you can add the fact that Northern Europe's built environment is not at all fit for purpose. Post-COVID, our housing stock is overused as people spend more time working from home, thus rendering vast areas of workspace underused or, worse, vacant. And, yet, in the UK, we are still building new office blocks. In

DOI: 10.4324/9781032665559-2
This chapter has been made available under a CC-BY-NC-ND 4.0 license.

addition, for the most part, the built environment is very poorly insulated, rendering it often uncomfortably hot in the summer and cold in the winter and, lately, too expensive to occupy. Finally, almost none of the UK's built environment is 'climate resilient', that is, suitable for a world with accelerating climate change. As I write this introduction, over 8 million households in the UK are suffering 'fuel poverty', the definition of which is when you spend over 10% of your income on energy bills.

Resource security is a thing of the past, and although most humans on Planet Earth have been aware of this fact, it probably wasn't until the early 2020s that the rich 'Global North' nations began to realise the same. For the last 40 years or so, resources have been so plentiful and comparatively cheap when compared to labour costs that building contractors got into the habit of over-ordering stuff just to make sure that their contractors didn't run out of things to do on site. Because of this practice, which includes allowing for 10% wastage when ordering tiling, bricks, ready-mix concrete, plaster, etc., it is common practice for skips to be filled with perfectly useful materials.

Back in 2012, there was a striking statistic published by Waste and Resources Action Programme (WRAP), founded by the UK Government, that stated that for every five houses constructed in the UK, one house-worth of material was being sent to landfill or incineration. This material included demolition debris, spoil from site, and most alarmingly, surplus new materials and components. What was maddening about this is that these waste streams were being factored into the cost of developing buildings so that they became, in effect, an invisible consequence of constructing buildings. Supposedly hard-nosed developers were wasting their own money sustaining this wasteful practice without realising it. However, this didn't go unnoticed across the industry. In 2012, the embryonic Superuse Studios (then 2012 Architecten) set up with a vision "to reduce the use of natural resources through innovation and clear design". Even earlier, Rotor were formed in 2006, describing themselves as a "research and design practice that investigates the organisation of the material environment". This led to Rotor Deconstruction being set up in 2016 as a cooperative that "organises the reuse of construction materials through dismantling, processing, and trading of salvaged building components." This was also the year that I called a mini 'waste summit' at my studio near Brighton. Here, I met Cat Fletcher, who helped form FREEGLEUK, "an exchange for unwanted stuff", with over 3.5 million subscribers. We also met with Dr Ryan Woodard, a research fellow at the University of Brighton, who has been working in waste management research for more than 20 years, along with product designer and academic Nick Gant and Diana Lock from the environmental management consultancy Remade SouthEast, which was set up to encourage businesses to consume less. This meeting led us to the decision to look for ways of raising awareness of these issues by creating pedagogic tools and inclusive working methods to get academics and students to think about construction in ways that dramatically reduced the need for new materials and turned linear systems into circular ones. We needed to re-learn and teach new ways of practicing.

The University of Brighton's first manifestation of this new thinking was the design and construction of the world's first permanent public building made from over 55 tonnes of material other people had thrown away. Named the 'Brighton Waste House' by a journalist writing for *The Guardian* newspaper, it was actually not a house but a two-story teaching facility and live, ongoing research project based on campus at the University of Brighton, designed and constructed in partnership with over 360 students and apprentices. At the beginning of the project, Cat Fletcher found a phrase on the website Treehugger, one that we used as our slogan for the duration of the construction period (the project was completed in 2014, on budget and on time!)

and one that has served us all very well since: "There's no such thing as waste, just stuff in the wrong place!"

The Waste House did a very good job of drawing attention to the problems of over consumption and waste generated by the construction sector. So, in 2018, when Rotor contacted me to see if I would be involved in their bid for funding from Northwest Europe INTERREG to help create a cross-border project that would facilitate an increase in reclaimed building elements from a paltry 1% today, to 50% by 2032, I said "yes" pretty much immediately. I was very keen to be involved as they wanted me to focus predominately on developing the student-facing pedagogic output for this cross-nation project. In addition, Rotor had developed an unrivalled reputation in the field of building deconstruction and, crucially, for enabling the reintroduction of secondhand materials and components into the construction industry supply chain.

Rotor's bid for over €4 million was successful and the project was given the name 'Facilitating the circulation of reclaimed building elements' in Northwestern Europe, or FCRBE, for short. Another attraction for us at the University of Brighton (UoB), where I work part-time, was that one of the main remits of EU-funded INTERREG projects is that they encourage inclusive innovation and knowledge exchange in industry and academia, across international borders. So, I assembled a team for the University of Brighton that comprised many of my colleagues from that first 'waste summit' in 2012. In addition, Rotor assembled the following partners:

Salvo Ltd – UK-based digital marketplace for reclaimed building materials since 1991
Embuild – (formerly Confederation Construction) is the main business organisation, social partner, and representative body of the construction industry in Belgium
Buildwise – the statutory members of Buildwise (Belgian Building Research Institute) include more than 90,000 Belgian construction companies, mostly SMEs, in its primary business performing scientific and technical research for the benefit of its members
CSTB (Centre Scientifique et Technique du Bâtiment) – the French national organisation providing research and innovation, consultancy, testing, training, and certification services in the construction industry
Brussels Environment (BE) – the public administration responsible for environment and energy in the Brussels region
University of Brighton – Baker-Brown's academic base for over 25 years of research into sustainable design, re-use, designing-out waste, and closed-loop systems
Bellastock – an experimental architecture organisation whose work focusses on the valorisation of places and their material resources

Three further partners joined the project in 2022:

City of Utrecht – UTR is the local authority of Utrecht. It is involved in multiple (EU) networks and projects (INTERREG, URBACT) on circular economy, business-modelling, and circular construction
Luxembourg Institute of Science and Technology (LIST) – this group provides science-based support for environmental policies at national and EU level and gives regulatory support for RDI (Research Development and Innovation)
Delft University of Technology (TU–Delft) – the oldest and largest technical university in the Netherlands, includes 8 faculties offering 16 bachelor's and more than 30 master's programmes for more than 25,000 students.

In addition to this extensive list of project partners, we had a group of industry experts with expertise in fields as varied as whole life carbon counting, re-use, social and climate justice, circular economy and closed-loop systems, systems thinking, and inclusive design. Known as our 'Associate Partners', they were taken from across the NWE INTERREG region, which comprises six Northern European countries, namely Belgium, France, UK, Netherlands, Luxemburg, and Ireland, and the participants acted as critical friends for the core team. The FCRBE project team also regularly invited members from the NWE INTERREG CHARM (Circular Housing Asset Renovation & Management) project who are investigating similar and complimentary themes to our own FCRBE project.

As stated before, my primary role was to design and curate the pedagogic output for this five-year project. On behalf of the University of Brighton, I assembled a team of eight academics who were charged with the task of developing an international summer school based in Brighton and Hove on the south coast of the UK. The event needed to directly address the FCRBE research focus, namely, the deconstruction of buildings instead of demolition and the re-use of secondhand materials in new build projects. So, in partnership with Rotor and Bellastock from Paris, we created five 'themes' that would be addressed during the summer school. These were the raw, the useless, the by-product, the hybrid, and the offcut.

Summer school themes needed to compliment/respond to the main question the FCRBE project asked – namely, that of considering re-use in the construction sector to avoid the current excessive and environmentally destructive use of raw materials and creation of massive flows of waste to landfill and incineration. As such, all ideas around the themes to be pursued, the design and curation of the event content, parallel programmes (such as the public lectures and lunchtime debates), and the selection of summer school team leaders and students were peer-reviewed at regular intervals by all FCRBE partners.

We planned a two-week summer school for August 2020. Based in an empty secondary school in Brighton, we named the event 'The School of Re-Construction' and anticipated having ten different teams of eight students each assisted by two team leaders who were academics or practitioners (or both) focussed on re-use in the construction sector. We invited students from universities offering engineering, design, and architecture courses across the six-nation region. Students were asked to submit a curriculum vitae with a 500-word 'motivational statement' in response to the five research themes. We accepted 80 students from over 150 submissions. They came from all six countries in the NWE INTERREG region. Students were studying at both undergraduate and post-graduate levels. It should be noted that as well as enormous support from INTERREG, the University of Brighton, and all FCRBE partners, the 'School of Re-Construction' (SoR-C) team worked very closely on the development of the summer school with officers from Brighton and Hove City Council (BHCC), who were, at the same time, developing their own 'Circular Economy Route Map'. BHCC officers were very keen to be involved with the summer school, and it was at their suggestion that we based the event at one of their 'mothballed' schools. In addition, and to my utter amazement, BHCC suggested that they would commission the deconstruction of one of their buildings instead of demolishing it. The idea was that the de-constructed parts of this building would be stacked in the playground we were using for the summer school and that our ten teams of students would then spend two weeks appraising this material, shining a light on its numerous potentials. And there was one last point. BHCC commissioned the construction of its own projects, and it has its own circular economy route map. As a consequence, the council members were keen to test the viability of closed-loop systems in their own 'live' projects, so at the end of the summer school, they were preparing to disperse the de-constructed material and put it into their supply chain to test if their contractors were prepared to work with it.

Unfortunately, the COVID-19 pandemic put paid to all the plans to host a face-to-face School of Re-Construction in 2020. The 80+ students who had been invited to attend the summer school were offered the opportunity to take part in the event in 2021. However, in early 2021, it became apparent that, due to COVID-19, it would still not be possible to host a face-to-face summer school as originally planned. And so, the FCRBE project team agreed to host a digital version of the summer school instead. Although the inability to host a face-to-face event was extremely disappointing, it soon became apparent that a digital event offered several attractive opportunities that would not be feasible if the event was site-based and face-to-face. So, we were able to invite students, not just from the NWE INTERREG region, but from across the globe. Consequently, we had applicants from countries as far apart as Columbia, Canada, Ireland, UK, Europe, Jordan, UAE, India, and China. We were also able to invite expressions of interest from academics and practitioners keen to take part as team leaders for what now became our International School of Re-Construction. Like the student participants, prospective team leaders were asked to submit curriculum vitae and motivational statements that included a 500-word 'statement of intent' describing their response to one or more of our five themes, together with an idea of a brief they would set students during the two-week digital summer school. As with the student participants, we received team leader applicants from across the globe. They included Prof. Folke Köbberling, who is an artist and Head of the Institute for Architecture-Related Art at the University of Braunschweig, Germany; Prof. Graeme Brooker, Head of interior design at the Royal College of Art, UK; Filipa Oliveira, an architect and coordinator from ACAN; Nicole Maurer from Maurer United Architects; Michaël Ghyoot and Sophie Boone of Rotor; and Jonny Pugh of Flores & Prats Architects.

The digital School of Re-Construction ran from 02–13 August 2021. Twenty-four academics and practitioners ran eleven separate teams of students. In addition, we had eight public lectures and debates given by internationally renowned experts in the field of re-use and designing-out-waste. These events were attended by over 110 participants of the summer school, as well as a further 250 external visitors. All recordings of the live events, together with details of the team leaders and the themes they pursued, can be seen on the International School of Re-Construction website (https://blogs.brighton.ac.uk/schoolofreconstruction/).

We believe that the process of developing the shape and content of the summer school resulted in an event of the highest calibre. Some of the most accomplished thought-leaders in the world of low-carbon, climate-resilient design worked with a cohort of very talented and motivated students, asking some of the most pertinent questions of our time whilst visioning projects committed to considering how humans can live in harmony with Planet Earth. However, to combat the existential challenges that the climate and ecological emergency presents us all, we need to stop working in silos and share best practices and, most importantly, be just as transparent about our failures and limitations. So, it was particularly exciting when we had the opportunity to make another bid for NWE INTERREG funding to support the writing and publishing of a book capturing the 11 teams' work over those two weeks in August 2021. With the help of Prof. Graeme Brooker, together with the approval of all FCRBE partners, we secured further funding in January 2022 to facilitate the writing of this book which we have now titled *Re-Use Pedagogies: The International School of Re-Construction*. Over the forthcoming pages, you can read chapters written by each of our team leaders giving detailed accounts of their team's experience participating in the summer school. In addition, we have two chapters considering what we called at the time 'the mega-themes' that were discussed over lunchtime debates hosted during the summer school. The final part of the book is dedicated to reflections on emerging re-use practice and pedagogy with chapters written by Lionel Devlieger and Maarten Gielen of Rotor and Professor Elma Durmišević.

Both Graeme and I would like to take this opportunity to thank our wonderful Editors Fran Ford and Hannah Studd at Routledge as well as our fantastic Copy Editor Kate Fornadel. And finally, we would not have been able to deliver the original International School of Re-Construction let alone this book without the dedication, rigour and support of our Project Manager from the University of Brighton Siobhan O'Dowd. Thank you Siobhan!

We hope that you enjoy the book.

THE ALLOTMENT

A test-bed for the circular economy?

Lucy-Ann Gilbert

Preamble

To make clear from the outset, through this chapter, you will intentionally be led through a story, an anecdotal account of my personal experience at my allotment – with some embellishments, of course. And, although, at times, this story may seem to deviate from the central topic, it is this very nature that is, in fact, for me a truer reflection of the circular economy – unpredictable and centred around human connection to a specific place and time.

Beyond this, I also hope that this method of writing will serve as an accessible way into understanding some of the principles of the circular economy, one that allows you, the reader, to reflect upon and learn from your own experiences.

So, without further ado, on to the formal introduction.

Introduction

Things will and do come full circle, it just depends how – what is destroyed in the path or, instead, reinstated? Having started on the School of Re-Construction project with Duncan Baker-Brown, Anthony Roberts and Siobhan O'Dowd back in 2019, we could not have foreseen the obstacles that would arise along the way. Despite being thrown off track for part of the process, I am now here to share with you a personal project that emphasises many of the core values set out for the summer school. At this point, I wish to be explicit. For me, circular economy principles extend beyond a simple loop of *reduce, reuse, recycle, and remanufacture*; they, instead, can be viewed as powerful tools for making connections and relationships to our built and lived environments – this is where I shall begin.

Throughout, I will be making connections to broader themes and, as alluded to before, may sometimes appear to meander off the beaten track, but, rest assured, there is *method to my madness*.

Firstly, although in this context I may be seen as an educator, spatial designer and allotmenteer – I am very much coming at this from the position of 'active agent' in the world – which we all have our part to play. As such, rather than the focus being on pedagogic design practice and how

DOI: 10.4324/9781032665559-3
This chapter has been made available under a CC-BY-NC-ND 4.0 license.

we can embed circular economy principles into a design studio setting, I, instead, wish to share my much-cherished allotment site for this enquiry and how it may serve as a useful metaphor for the circular economy and beyond – after all, what better way than to practice what you preach!

Setting the scene

To set the scene, the School of Re-Construction is part of a much wider project, as the summer school within the INTERREG research project titled '*FCRBE – Facilitating the circulation of reclaimed building elements*.'[1] Extending beyond this, as hosts at the University of Brighton, our initial aim was very much to engage with the local context, not only deconstructing and reconstructing from live sites in the city but, in turn, also feeding into the circular economy route map that was in the process of being developed by the local Brighton and Hove City Council.

Although the live project element became digital due to the dreaded 'C' word, the relationship to the city was still very much at the fore. Developed over several years, the 'Brighton and Hove Circular Economy Route Map: a sustainable green growth strategy for the city' has now been disseminated into the public realm, and as they state, '*We will initially concentrate on two key sectors for the city: the built environment and Food & Drink, before expanding to look at other areas*.'[2]

This brings me nicely onto the case study I wish to use – my allotment, also known as "Plot 132a". With an allotment being described as '*a plot of land rented by an individual for growing vegetables or flowers*.'[3]

"What has your allotment got to do with the circular economy?" you may ask. For me, however, the connection is clear as day. Upon reading the plans for our city's route map – focussing on our built environment and food – the correlation became ever clearer with the allotment providing the ideal microcosm where, to some extent, these principles are already in action. With the many built structures often utilising material otherwise considered 'redundant' and food grown used not only to feed ourselves but surplus shared with family, friends, neighbours and local food banks and any food deemed unsuitable is recycled as compost for the following year – *there's no waste here*! As such, I began to contemplate what could be learnt from the humble allotment?

A space arguably for growing but, as so poignantly presented in the study by Brighton and Hove Allotment Federation, a place that offers so much more – from '*biodiversity, storing carbon, supporting bees, improving physical and mental health, reducing loneliness*' – all of which greatly benefit our communities whilst saving our council money in ways that may not be explicitly evident.[4]

In a similar vein, although, to some extent, my plot is about facilitating the growing of produce, it is also significantly richer in its ability to draw together many different aspects of one's life and is arguably more emblematic of a way of engaging with the world than simple cause-and-effect outputs such as plant, water, grow, harvest, eat, compost and repeat.

To give some examples, on a personal level, I find spending time at the allotment is a great way to slow down and engage in the present moment – even when I try to go with a plan, often, upon arriving, it changes due to more immediate needs that come from working with our external environment and the seasons. It is largely due to this transient and unpredictable nature that it somehow feels safer to experiment, opening up possibilities based on the changeable factors that arise – with all these parameters at play **could allotments become the ideal test-beds for the circular economy?** Providing safe places for experimentation as well as sites to utilise some of the less desirable surplus from demolition sites. And to some extent, are they already – in an implicit way at least doing this? From my experience, there is no lack of supply or demand. The issue is one of timing, communication, storage and transportation – therefore, could our circular economy route maps begin to build on such relationships more explicitly from the outset?

It is these questions that I wish to ponder throughout this chapter starting with my own allotment plot, how it evolved and my collaboration with others then leading to the bigger picture and how this learning may apply to broader contexts. Through my story, I hope along the way it will begin to reveal the reciprocal relationships and humanness necessary in the making and exchange of such an economy.

My allotment story

I am not quite sure how it happened, but after putting my name on what I was told was "a very long waiting list," approximately three months later, on October 28, 2018, I became the proud owner of my very own allotment plot. As you may suspect, this was much faster than I had anticipated, and with little to no experience of really growing anything at this stage, I cannot deny I was pretty overwhelmed. I would include a personal photo here but it was all rather overgrown so difficult to capture – instead, Figure 0.1 shows a Google map view demarcating the boundary of my plot:

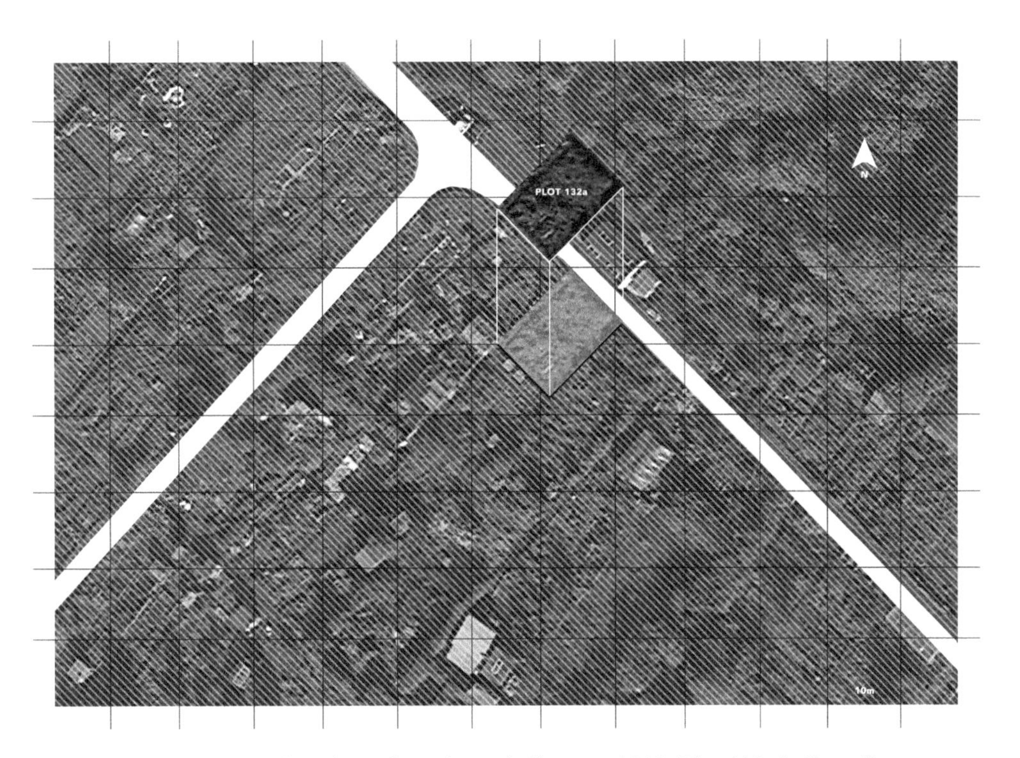

FIGURE 0.1 Google Map plan view of Natal Road allotment 2018 (Plot 132a indicated)

Welcome to Plot 132a – mainly grass with some hawthorn and damson to the perimeter and raspberries in the centre. I later found out from my allotment neighbour that this plot had been owned by someone called "Roy" for many years prior, and a few tenants had tried but failed to get it going again – this added somewhat to my fear of failure but, at the same time, fuelled my desire to meet the challenge.

I had decided from the start that rather than jumping right in, (although this was contrary to the advice I had been given – just throw some seeds in, you'll get evicted if you don't grow in the first year!) I was going to take a slower more observational approach – partly due to my inexperience of growing but also coming from an interior architecture background I felt a deeper understanding of my specific context and the existing site was essential (as I later learnt in permaculture terms, this is also known as '*unthemed observation*' adding further validity to my approach[5]). And so I began, quite arbitrarily, spending time there, drinking tea and wandering around other plots to gain a sense of the possibilities – I learnt quite quickly allotments do have rules – primarily, '*the percentage of land to be cultivated at a minimum 75%* and *structures and trees not to exceed 2.5 m high*' – as I soon found out after receiving a warning that first summer for weeds and height of the (in my defence) *existing* trees. It was safe to say the council did not necessarily agree with my slower approach.[6]

If we travel back in time a little to my first winter at the allotment 2018–19, I became increasingly aware of my need for shelter, and – despite the purpose of the allotment being to 'grow' – having had experience both designing and building, I really began my allotment journey here – with the shed (or 'cabin' as some of my friends prefer to call it).

For those who have not had the pleasure of exploring an allotment site, there is a real sense of wonder and magic that comes from the ad hoc methods of construction arising from using what is available, and this essence is something I did not want to lose. At this stage, however, beyond the odd thing growing on the land itself, I had only discovered a few pallets and some large rocks hidden in the undergrowth, meaning my search had to broaden which led me to Freegle, '*a UK organisation that aims to increase reuse and reduce landfill by offering a free Internet-based service where people can give away and ask for things that would otherwise be thrown away,*' founded by the local waste legend, Cat Fletcher.[7]

Through this platform, I was able to source free timber; however, I soon found the search became increasingly challenging due the quality and quantity of material people are giving away (often already rotting and/or small awkward offcuts). This was made more complicated by the fact I can't drive (maybe one day); therefore, by the time I was able to organise someone to help with collection, the material would often be gone.

Unsure how to proceed in what felt like an endless search, I found the material was under my nose all along.

The Lectern Pub

One November morning in 2018 on my way to work, now primed to look for potential material sources wherever I went, the penny finally dropped – the Big Build – '*the name our students have given to the project that is transforming our Moulsecoomb campus.*'[8]

With the Big Build came mass demolition, and one of these sites was the Lectern Pub described as '*Lewes Road's finest pub and quite possibly the best student boozer in town.*'[9] I wouldn't quite go as far to say it was the 'best' but it did have its place in the student culture of the time – here I should probably add that not only have I taught but I am also former alumni of our BA interior architecture course at the University of Brighton.

Why does this all matter? Well, for me, this is where everything felt like it began to fall into place, in this serendipitous moment – with the re-using of material. It is always more than just stuff to build with and, instead, holds a form of material meaning and value that comes from the cultural context and narrative of the existing site in which it derived. Put yourself in my shoes and imagine if someone were to have told me 15 years ago that I would be writing a book chapter about circular economy and the shed I had built at my allotment was re-using elements of the old Lectern Pub I had formerly frequented – I would have said, "You're crazy!" It would have seemed too surreal and farfetched to even contemplate. Alas, it is true.

It is hard to pinpoint exactly what the significance is. I feel it sits somewhere in the uncanny where '*the difference between familiar things that delight us and familiar things that terrify us start to make sense.*'[10]

In my case, the delight being in the re-use of material from a building I had inhabited many years before alongside the terrifying reality of the Big Build and why I had access to this material in the first place. Here I wish to point out that, in essence, I do not disagree with the Big Build (providing much needed housing and additional social, work and university spaces). However, what I do find puzzling is how this has been realised – using largely outdated steel and concrete construction that appears to be evident in many of the new builds appropriating previously undeveloped sites across the city. I do not want to assume, but I do wonder how and why this all happened so suddenly, just before the local circular economy route map was implemented? And at what cost – not only monetarily but also environmentally and socially? How could it have been done differently if circular economy principles were considered from the outset?

Anyway, I digress

With my eye firmly set on the Lectern demolition site, all I had to do now was somehow gain access to this material. Time was very much of the essence. After patiently scouting, I knew that if I did not jump at the opportunity, this material would be gone before I knew it. A rather strange series of events proceeded, but you'll be pleased to learn that my desire became a reality.

One morning, after eyeing up this timber for about a week, I finally plucked up the courage to simply ask one of the workers what was going to happen with all the demolition material to which they didn't really seem to know beyond "it'll get dumped." As it was clear it wasn't useful or deemed valuable to them, I then very quickly asked if I could have some, pointing ahead at a large pile of what I believed to be floorboards. They understandably were reluctant to just say yes, but this did not deter me, so I arranged to meet with the site manager the next day. After explaining my intention to build my allotment shed with it, I could tell they thought I was little kookie, but they agreed, nonetheless, and we got down to logistics of how much I wanted and when – believe it or not, early one morning the next week, we met, loaded a van and drove the material up to my allotment (only approximately 0.3 miles away – Figure 0.2) somehow it now all seemed so easy!

Once I had the material, the real work began – a labour of love as I de-nailed and cleaned approximately 125 m lengths of timber, taking around 20 hours – therapeutic, some might say. It was early February by the time I was done and again spending time on my plot now with the material, but plans soon evolved.

As most of the year's growth had died back, I used these winter months to clear the site, again sitting in various spots sipping tea – I finally reached my destination, the location of the shed – Figure 0.3 panoramic photo taken standing at the southernmost point facing due north towards the Hollingbury Hillfort, the spot provided a full view over the plot and the hills beyond.

With my material and site located, I was finally ready to move on to the next phase – the design.

FIGURE 0.2 Google Map view of site for "Big Build" and Natal Road allotment 2018 (Lectern Pub and Plot 132a indicated)

FIGURE 0.3 Panoramic photo of allotment plot (March 2019)

The design (and build)

My initial brief for the shed evolved from the time I had spent at the allotment and was to provide me with the necessities I had been craving, including:

- shelter both from wind and rain but also a source of shade on sunnier days
- a place to store some basic tools for building, growing, cooking, eating and preserving
- being den-like in scale because I wanted it to feel like a space to retreat
- a place to sit and lay if the mood struck
- a way to harvest rain water for my crops
- re-using material where possible and, if new, sourcing locally to create minimal waste by reducing/utilising offcuts

With these design parameters in mind, I began sketching, working firstly in plan refining the orientation and size of the base structure but quickly realising that to move beyond this I needed to work more closely with the material I already had. This was something new to me as often when designing I would design then later source and specify the material and quantity for the job.

Working with a known material and quantity in this way became key in order to minimise both waste produced and, importantly, the number of cuts made – as it's an allotment site without electricity, the labour time and effort also became a critical factor. Although it presented some new challenges, again with it came a slower way of working and a need to find innovative solutions to things that in other cases would not even be a problem.

I had toyed with the idea of making a physical model as I would typically; however, in this case, with the real material at hand, I set about surveying, sorting and creating an inventory of the reclaimed timber from the Lectern Pub.

After this (unusually for me), I jumped into creating a 3D digital model alongside hand-drawing and working with the material on site as this gave me a more rapid and precise way of working with the components I had. Finding myself in a kind of chicken-and-egg situation alongside this drawing and modelling process, I needed to search for any other materials I required – due to the challenges in finding this timber, I decided for the main frame I would work with local suppliers but for these standardised components to remain at full size where possible to reduce waste and the need to cut!

This evolved into quite a simple yet impactful form, as can be seen in Figure 0.4, that made the most not only of these measured constraints but importantly also the view and roof surface for rain water while creating the intimate space I so desired.

Through this process of drawing, modelling, testing on site and sourcing materials, by April 2019, I had my tools at the ready and batteries charged and was finally ready to build – various stages of which are shown in Figure 0.5.

Collaborate (and listen)

For anyone with making experience working at any scale, we always feel we could do with an extra hand – luckily, I had some "willing" helpers to call on – special shout out to Brittany Wardle, Samantha Lynch and Rob Vinall and to my dad (Adam Gilbert), my allotment neighbours and the various suppliers, in particular those demolishing the Lectern – it wouldn't have been possible without you!

I do not wish to push a purely feminist agenda here, but in an industry dominated by men, I feel it is incredibly important to bring attention to the need for collaboration in the success of applying circular economy principles and our ability to tackle the climate crisis more broadly. It is not about master, ego, leader but the collective sharing of knowledge and expertise across a range of fields – after all, no (wo)m(x)n is an island – and this includes *all* of us in our communities.

It is here I would also like to make very special mention to the late and great Julia Dwyer (former tutor and a founding member of feminist collective Matrix[11]) and to Joel Bird (shed builder and collaborator[12]) without which I may never have gained the confidence to build in the way I have – a skill I, too, hope I am able to share with others throughout my lifetime.

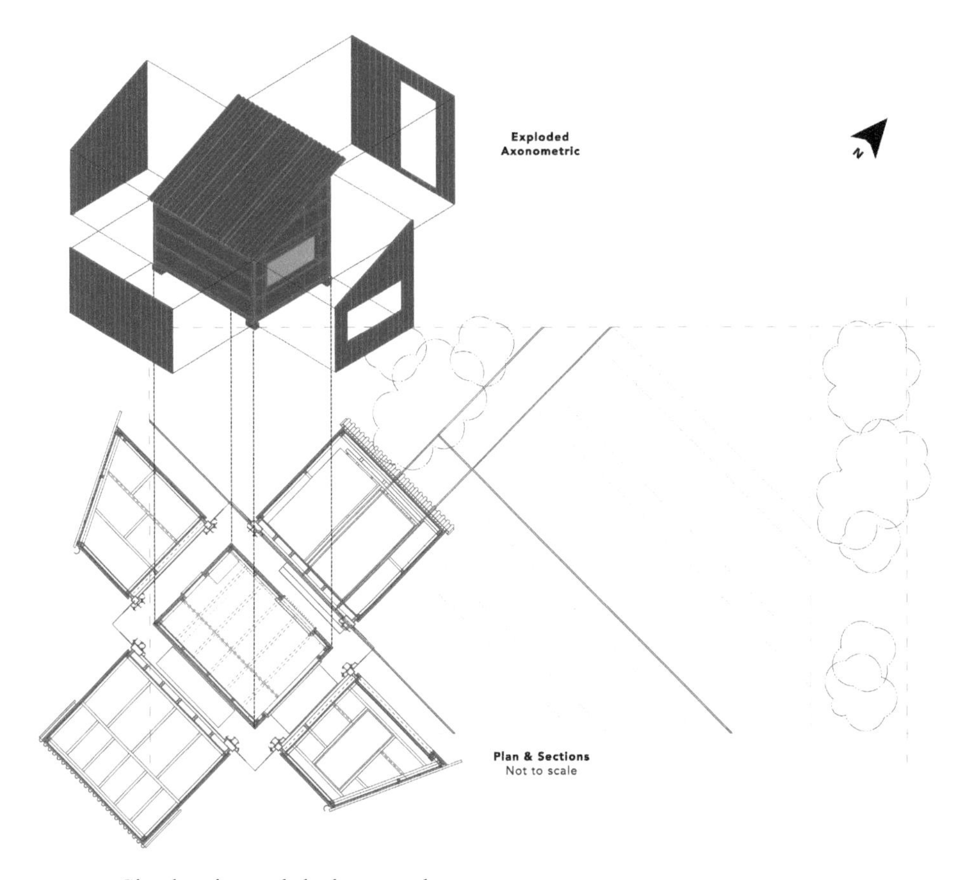

FIGURE 0.4 Site drawing and shed proposal

I shall continue . . .

At the outset, it had not been my conscious intention for the level of collaboration that ensued. After all, it is "only a humble allotment shed," but upon reflection, it has become much more than that. On a personal level, the allotment not only serves as a space of my own but also acts as a useful reminder for the holistic ways in which we can live *with* our environment.

We do not have to have grand master plans to realise the potential the circular economy can offer – in fact, one might argue that by reducing the scale and speed of our ambitions, it could help us to observe and learn from the wider systems at play. By this I mean that although our intentions can be "good," if we do not take the time to notice the bigger picture, we often end up only treating the symptom and not the cause.

To put it bluntly, '*fuck good intentions*' as quoted by Ben Sweeting in his talk at the 2022 Relating Systems Thinking and Design (RSD11) symposium titled *Architectural Roots of Ecological Crisis*.[13]

In the face of the climate crisis, for greater chance of success, we must slow down and take a moment to pause. As Anupama Kundoo so pointedly states, '*We feel like we don't have the time, so we don't take the time to think . . . our time too is a wasted resource*.'[14]

Ask yourself, what is it that we are really doing here with our precious time? Yes, there is procreation and evolution of the species but at the core of our existence is relationships – to one another and the world around us.

FIGURE 0.5 The shed build process (2018–19)

These relationships form an extremely complex system that is almost unfathomable and may be why we often appear to fall into the trap of narrowing our focus of identifying problems and providing "solutions." Although the speed and rate in which we act can have grave consequences as the knock-on effect ripples and spreads in ways we may have not yet perceived. I do not wish to bring a pessimistic view that prevents us from moving forward instead to ensure our "good intentions" are reversible, if needed.

Again, back to my allotment, where, for me, lies a prime example of these wider interconnected relationships between the seasons, land, animals, people, infrastructure, material, growth, food, decay (Figure 0.6) – a microcosm where our natural and built environments merge – and a place that each year slowly evolves, regrows and, to some extent, will continue to do so with minimal human intervention all started by planting "the seed."

If you still haven't yet had the pleasure to wander around an allotment site (if you can – do!), it is quite a spectacular and surreal experience seeing these various worlds at play – the homogeneity of our "real world" appears to be somewhat liberated here due, in part, to the transient and ad hoc nature of each individual plot and its evolution – again maybe useful metaphors for a more circular approach – not everything has to be or should remain the same.

It goes back to the idea that there isn't a one-size-fits-all approach and, therefore, by moving away from the repetitive and homogenous model that often gets applied to our built environment, not only could we be living in spaces that are more akin to our specific context but also testing multiple systems while spreading the demand on our world's precious resources – as my and many mums would say, "Everything in moderation!"

FIGURE 0.6 Allotment plot through the seasons (2020–21)

Food (for thought)

Linking back to the Brighton Circular Economy Route Map with a focus on construction but also food, this notion of "sameness" in very much evident in both these areas – construction materials used and the ingredients we consume.

Many of us (myself included) are probably guilty of blindly food shopping and, for the most part, selecting similar ingredients from our supermarket's shelves week in, week out. Although the choices often seem endless, what if we took a more conscious approach to local and seasonal produce? In the same way, what if our built environment stopped mimicking the "new" efficient design-and-build architecture of its time and, instead, utilised the local materials available including deconstruction sites as a resource?

At first, it may appear limiting, but in reality, would we not be more adventurous with our diets being more varied across the year (not to mention even tastier) and our buildings more personal and site-specific and holding more contextual richness?

For this to happen, it feels a shift in mindset is required, one that's values are not driven by money and efficiency as there is no denying that this approach would take longer and potentially cost more – but at what *real* cost? What are the policies and incentives that need to shift? And this begs the question, what is the actual goal? If it is about building a sustainable future for ourselves and subsequent generations, then, surely, we cannot continue to be so short-sighted.

As addressed by Anupama Kundoo in her seminal lecture '*Rethinking Urban Materiality: Time as a Resource*' at the Institute of Architecture Southern California in 2022, '*We are losing knowledge by not using what is around us both materials for construction and food.*' (See footnote 14).

Therefore, arguably, by not engaging more fully with our immediate environments, we are severing generations of knowledge which would be a travesty at large – again this points back to the essential need for collaboration and skill-sharing within our communities.

A slight deviation, but to further bridge this relationship of collaboration through construction and food, Brittany Wardle (a.k.a. Britt), one of the friends who helped me with the shed build, was, at the time, head chef of local Brighton restaurant the Flint House. Why is this significant? Well, as Britt had no real prior building experience and I no formal cooking training, we felt this an ideal exchange of skills – her supporting me with the build and then later using produce from the allotment (Figure 0.7) to provide us with the local and seasonal ingredients to begin testing combinations for new dishes at the restaurant.

The brief was simple. Much like a famous cooking show of our youth, we had to use as many of the ingredients as possible with access to essential items from the pantry – process and finished dishes shown in Figure 0.8 . . . Not only did I learn some tricks of the trade and make delicious food but two of the dishes even made the menu at the Flint House (marinated beetroot salad with miso and cavatelli pesto pasta with tomato)!

For me, this collaboration is a prime example of how the allotment can begin to open up possibilities in relation to circular economies, by making connections to our wider environment. With this in mind, what are the other connections that begin to shape and inform the bigger picture?

Conclusion (the *bigger* picture)

Let's end where we began – come full circle if you will – and take this moment to pause and reflect upon what has preceded. Through re-reading this "story," what themes are emerging and how may these help us understand the circular economy in relation to the allotment and the wider context?

FIGURE 0.7 Selection of produce grown at allotment (September 2022)

Time, speed, scale; slowing down, being present, observing; site specific, contextual, narrative-based, holistic and collaborative; transient, unpredictable, uncanny, ad hoc and serendipitous – to name a few . . .

These themes reaffirm that the circular economy is not a simple loop of *reduce, reuse, recycle, and remanufacture* but has many overlapping systems at play.

Taking this a step further, by mapping the *bigger picture* in relation to circular economy, the allotment shed and these wider systems (see Figure 0.9) we begin to reveal this ripple effect and how the circular economy touches upon much more than may initially be assumed.

Arguably, although this may be where it often begins – with the demolition or removal of material – the richness that ensues would not even be possible without the human interactions that arise within specific places and times. As this demonstrates, the wider systems at play are far-reaching and ultimately continue to grow as new connections are made – as demonstrated

FIGURE 0.8 Recipe experiments and skill-sharing (September 2022)

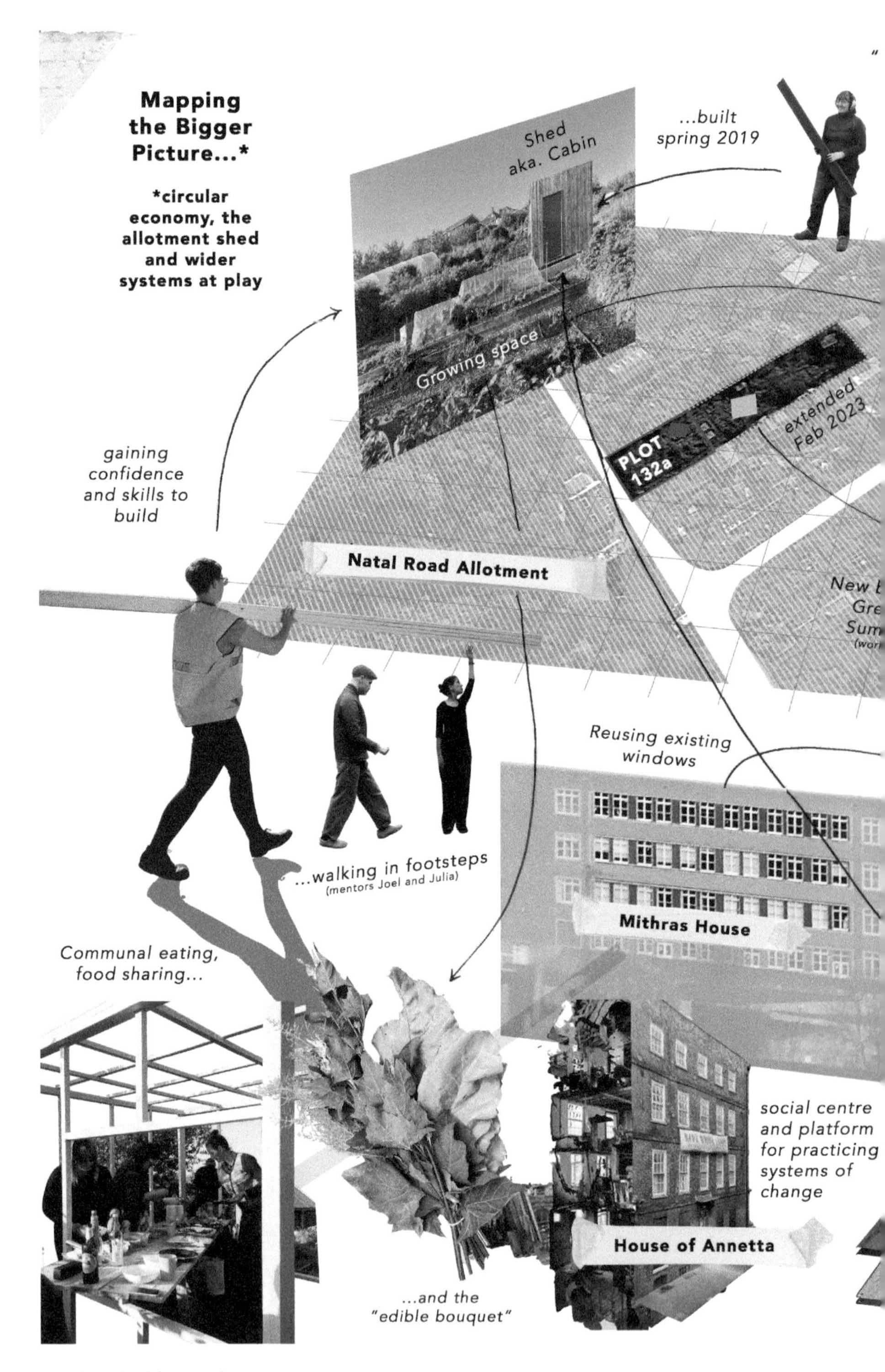

FIGURE 0.9 Mapping the bigger picture

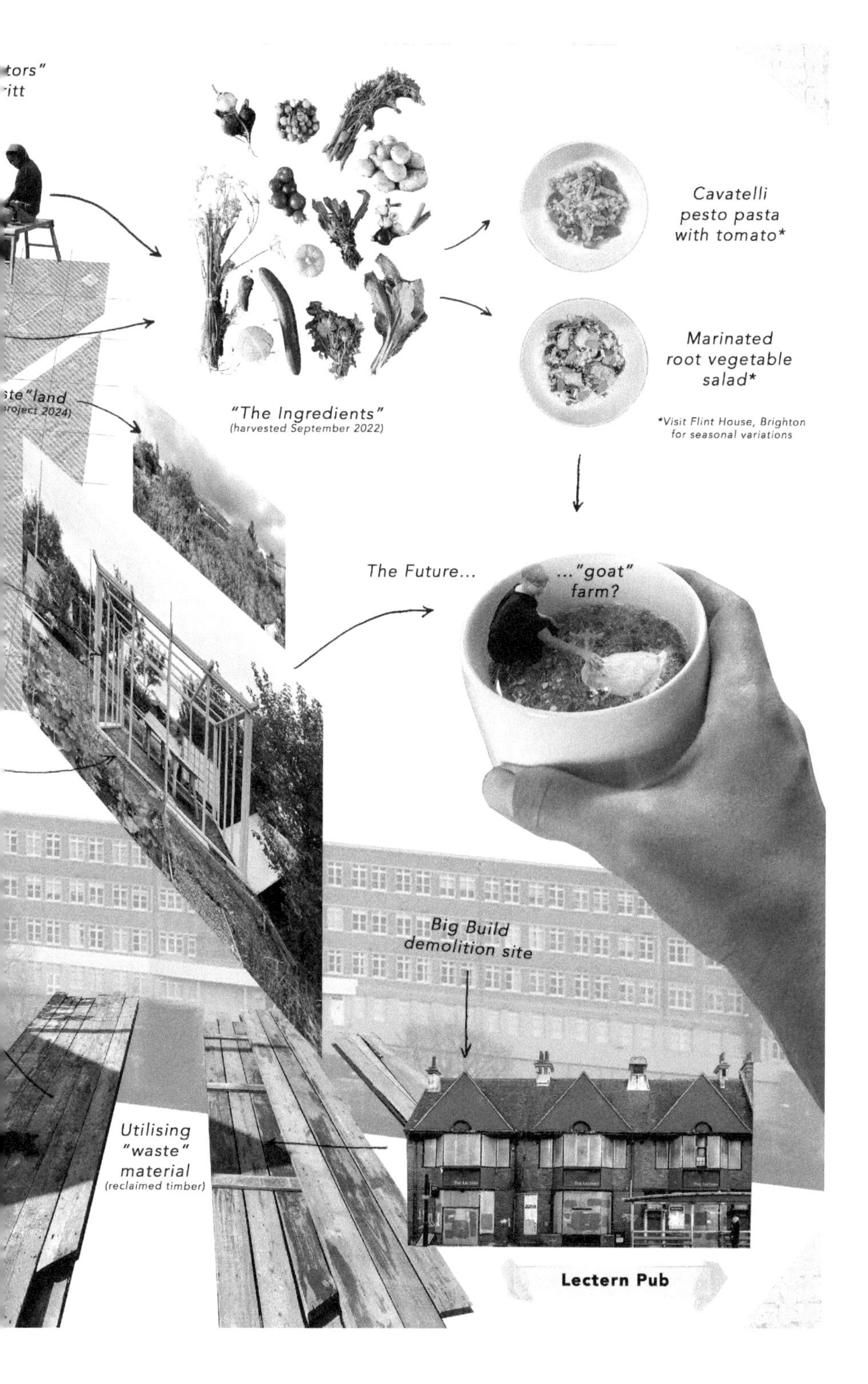

tors"
itt
ste"land
roject 2024)
"The Ingredients"
(harvested September 2022)
Cavatelli pesto pasta with tomato*
Marinated root vegetable salad*
*Visit Flint House, Brighton for seasonal variations
The Future...
..."goat" farm?
Big Build demolition site
Utilising "waste" material
(reclaimed timber)
Lectern Pub

here: from the extension of my allotment plot in February 2023, coinciding with the removal of the windows at the nearby university building at Mithras House, and, in turn, building of a new greenhouse; to the future community build project on the former *"Waste" land* at the Natal Road allotment site; and as far afield as collaborations and meals shared at *House of Annetta*[15], a social centre and platform for practicing systems of change near Brick Lane London – all stories for another time! The point is, if I were to do this again next year, who knows how things will have evolved?

In short, yes, I built my shed re-using parts of the former Lectern Pub, but the true value is so much more than the physical material that otherwise would have been wasted – from the skill-sharing through to collaborative building and cooking, communal eating, and numerous life-affirming discussions whilst also helping our planet – it's a win-win!

So, what can be learnt from my story?

Well, if I had to share one lesson, it would be the need to be open to the possibilities – possibilities of what you may find, who you may meet, and ultimately how things could unfold. It is unpredictable, for sure, but all the more fruitful as a consequence.

Through even these small interactions stemming from my allotment, the accretive nature of circular economy principles can be felt – opening up further possibilities for exchange, one that keeps on growing in many forms and guises.

If I had to answer the question – could allotments become the ideal test-beds for the circular economy? I would say a definite yes (but I am biased). For others, it may manifest in a different place or in a different form, but ultimately, I feel taking a slower approach, not driven by purely monetary gain, is essential to the success of such an economy. One that is rich in poetry and meaning through the relationships we form with one another and the world around us. This may require a change of mindset as we open ourselves up to the possibility that there are other ways of doing things – so embrace the uncertainty and get stuck in!

Apologies if you were hoping for a more concrete solution, but the reality is we will need to pose more questions than answers – so I will ask you this, what is your next step towards a more circular economy*? And it is with that thought I will leave you to ponder . . .

**for those who know me, yes I may have bent the truth in places – like any good storyteller – but you will have to wait for the sequel to find out if I finally reach my ambition of owning some land, building a house, getting some goats, growing my own food, running a community kitchen and making pottery to serve said food . . . watch this space!*

Notes

1 University of Brighton: Blogs-Brighton. *School of Re-Construction*. Accessed 25 August 2022. https://blogs.brighton.ac.uk/schoolofreconstruction/

2 Brighton & Hove City Council. *Brighton & Hove Circular Economy Route Map 2020 to 2035*. Accessed 25 August 2022. www.brighton-hove.gov.uk/business-and-trade/brighton-hove-circular-economy-route-map-2020-2035

3 Oxford Dictionary Definition. *Allotment*. Accessed 25 August 2022. www.oxfordlearnersdictionaries.com/definition/english/allotment?q=allotment

4 Brighton and Hove Allotment Federation. *The Financial Value of the Benefits That Allotments Bring to the City*. Accessed 25 August 2022. www.bhaf.org.uk/content/about/issues/the-financial-value-benefits-of-allotments

5 Gehrels, Stephen. *Introduction to Permaculture: Unthemed Observation*. Brighton Permaculture Trust, 21 January 2023.
6 Brighton & Hove City Council. *Allotments Rule Book*. Accessed 25 August 2022. www.brighton-hove.gov.uk/allotments/allotments-rule-book
7 Freegle. *Freegle – Like Online Dating for Stuff*. Accessed 25 August 2022. www.ilovefreegle.org/
8 University of Brighton. *The Big Build*. Accessed 25 August 2022. www.brighton.ac.uk/bigbuild/about/index.aspx
9 What Pub. *Lectern*. Accessed 25 August 2022. https://whatpub.com/pubs/BRI/450/lectern-brighton
10 Oregon State University. *What Is the Uncanny? Definitions & Examples*. Accessed 25 August 2022. https://liberalarts.oregonstate.edu/wlf/what-uncanny
11 Spatial Agency. *Matrix Feminist Design Collective: Julia Dwyer*. Accessed 5 July 2023. www.spatialagency.net/database/matrix.feminist.design.co-operative
12 The Shed Builder. *Joel Bird*. Accessed 5 July 2023. https://theshedbuilder.co.uk/
13 Sweeting, Ben. *Architectural Roots of Ecological Crisis*. RSD11 Symposium, University of Brighton, 16 October 2022.
14 Kundoo, Anupama. *Rethinking Urban Materiality: Time as a Resource*. Sci-arc, Institute of Architecture: Southern California, filmed 9 March 2022. Video of lecture, 1:05:45. www.youtube.com/watch?v=pZo0gIG63cw
15 House of Annetta. *About*. Accessed 1 August 2023. https://houseofannetta.org/About

PART II

The mega-themes

1

BUILD LIFEBOATS, NOT COFFINS

Reimagining architectural education for a Just Transition

Scott McAulay

During 2021's School of Re-Construction, I was invited to answer a question with a provocation: Is architectural education preparing students for the future as it will be shaped by climate breakdown? The succinct, unsettling answer was "not in the slightest"; this answer has barely changed since. My provocation could have simply been "no". However, to do the question justice, we must analyse the contemporary and historical relationships between architectural education and its climate inaction.

Essentially, the reality of planetary dysregulation is yet to be faced head-on, processed, and acted upon appropriately by traditional schools of architecture and those teaching within them. Climate change is yet to be taken seriously enough by the construction industry to spark any lasting, meaningful divergence from business-as-usual – to even try to leave the shameful "40% (of global carbon emissions) industry" moniker behind.

My project, the Anthropocene Architecture School, would not exist, nor would this book, had the architectural education system provided an education fit for purpose. It came as no surprise when the deafening response to 2018's Intergovernmental Panel on Climate Change Special Report on Global Warming of 1.5 °C was to continue as if it heralded no change for architecture at all – as if it was not terrifying and had not issued civilisation-scale deadlines. Just as there was little movement or change in architecture schools following the first Earth Day in 1970 or after the Club of Rome's Limits to Growth Report in 1972 and no unified action across institutions following the United Nation's Earth Summit in 1992, 2018's report also barely affected them, and courses continued as before.

A survey during the Architects Climate Action Network (ACAN)'s Climate Curriculum Campaign unearthed that 76.9% of students felt their course was failing to prepare them for future work.[1] It also broke news that 88.5% of students surveyed wanted to be tested on their ability to design sustainably but "aesthetics are valued more than sustainable design" where they studied. If those results are not damning enough, 2022's *Architects Journal*'s student survey revealed that around one in eight students graduates with little – if any – education on the adaptation and retrofit of existing buildings, and 4% of respondents were not getting any tuition in sustainability.[2] Given that 80% of the buildings that we shall be using in 2050 already exist[3] and

DOI: 10.4324/9781032665559-5
This chapter has been made available under a CC-BY-NC-ND 4.0 license.

that no country can hit a meaningful climate target without decarbonising its existing buildings, this is a destructive oversight.

Therefore, not only are the UK's architecture students not being equipped for the future, but many still meet with institutional climate denial and, by extension, negligence. Because of this, generations of students have been failed by their architectural education – fuelling huge knowledge gaps in an industry yet to grasp the scale of a planetary emergency. Now, things must change faster than traditional academic cycles deem possible.

This state of the architectural education system begins to demystify the wider architectural profession's ongoing resistance to change and reluctance to engage in climate action (unless there happens to be an opportunity for significant publicity, the chance to speak at a Conference of the Parties, or an associated awards dinner). This has been shaped, in no small part, by practitioners having received an education without sufficient teaching on how the world's systems – both natural and human-devised – shape buildings, the scale of their environmental impacts, and how construction intersects with climate change and, in turn, spatial justice. It is likely why the Anthropocene Architecture School lacks comparable contemporaries four years after launching during the Architecture Fringe and why there are still only student climate action groups at 13 of the UK's 61 schools of architecture.[4]

We must not overlook the fact that this is how architectural education is doing after the supercharging of climate activism in 2019 and the growth of ACAN from 2020 to 2021. Sustained activism in the wake of 2018's IPCC report led to incremental but significant policy wins – like inadequately far-in-the-future Net Zero carbon targets and declarations of a climate emergency from cities and countries alike. In the aftermath of the globally resonant School Strikes for Climate and the civil disobedience and truth-telling of Extinction Rebellion, architectural education remains obsolete in the face of climate breakdown and is by-and-large preparing to fail yet another generation of students. Despite such failures, there is yet to be a sustained movement of architecture students protesting these systemic failings – Architecture Education Declares arose, collected some signatures, then faded away – and that says more than words can.

Facing reality

In a just world, growing understanding of the climate crisis would have transformed architecture – its cultures, education, and practice; built space around us; and government policy worldwide. After all, as Naomi Klein put so powerfully: we are alive "at the last possible moment when changing course can mean saving lives – on a truly, unimaginable scale".[5] Yet, there has not been systemic, and infrastructural, changes sufficient to alter that trajectory. As you read this, our world is being driven – by conscious human decisions regularly putting profit before life itself – towards irreversible climate tipping points.[6]

The pathway humanity is currently on was never inevitable; governments have received warnings for over half a century. In fact, many in positions of power in the Global North actively choose not to let climate action, or decarbonisation, get in the way of neocolonial ideologies around "development," or shareholder profits – with especially loud shout-outs to Exxon Mobil, BP, and Shell for dishonourable contributions to delaying, derailing,[7] and backtracking on climate action.[8]

Temperatures are rising, floods are displacing people by the hundreds of thousands, and climate breakdown is omnipresent for millions of people, but onward "normality" rolls – no matter the cost, nor the size of the sacrifice zone. According to the laws of physics, our climate trajectory is not fixed,[9] meaning other futures are still possible: we have had sufficient technology to begin shifting to zero-carbon societies for more than a decade after all![10]

Learning in the Anthropocene

Despite 30 years of IPCC reports and 27 COPs, we are yet to see sufficient action or collaboration from policymakers to slow climate breakdown and save millions of lives. At COP26, the prime minister of Barbados, Mia Mottley, stressed that a 2 °C temperature rise is a "death sentence" for island nations like her own.[11] Even if today's policies were enacted in full, the world would warm by 2.2–3.4°C,[12] meaning many island nations would cease to exist.

The roots of today's ecological crisis lie in colonialism, extractivism, and imperialism, intensified by the burning of fossil fuels and the obscene consumption of the super-rich – billionaires having environmental impacts over 1,000,000 times greater than the average person.[13] The designed obsolescence of buildings and a wasteful construction industry,[14] with an appetite for demolition, are relatively unspoken about beyond construction "sustainability" spaces despite having an almost unparalleled climate impact.

Uncomfortably and dangerously, many climate "solutions" originate in Western-centric worldviews, influenced by deeply-rooted mythologies of human's separation from nature,[15] with colonial roots in mechanistic science championed during the European "Enlightenment"[16] and mythologies of technology,[17] and they often rely on technology that does not exist at sufficient scale. In their essay, *Buildings Designed for Life*, Amanda Sturgeon explains that buildings "embody our perception of nature as other – something to destroy or dominate",[18] and this colonially exported worldview has had catastrophic consequences.

Two things are certain: some kind of transition will occur as the climate crisis intensifies, and futures in which we equitably decarbonise civilisation look radically different than those of today. As costs of survival surge, many cannot afford to make time to reconcile with this future, let alone try imagining it otherwise. Education becomes more crucial because it can create space to do both, so curricula must be compassionately revolutionised to facilitate this. Architectural education could play a fundamental part in deepening our civilisation's capacity for a "Just Transition" – away from polluting fossil fuels, dualism, and consumerism and towards a more compassionate, circular, just, and regenerative future, without modernity's geographical and intergenerational sacrifice zones – should it learn to communicate, educate, and engage in new ways.

An inconvenient truth, and hindrance, is that those teaching in today's architecture schools were not prepared to facilitate such climate-literate, low-carbon learning by their own education. As Professor Susannah Hagan distils in *Revolution?*, "The past is doomed to be repeated if education repeats the past. Tutors trained in the old ways train students in the old ways", further exacerbated and exacerbating that architects still do not see the climate crisis as "the defining moral and intellectual challenge of our time".[19] Responsibility to transform architectural education lies with its educators and institutions; dominant notions and expectations of my generation and students shouldering that labour for them end right here.

To support students' learning for their unpredictable future, design studio projects must do more than request the articulation of learning outcomes and the ticking of boxes. Each one should be reimagined as an opportunity to offer glimpses into futures where better becomes possible. Instead of prescribing new buildings to "solve" hypothetical problems, students should be encouraged to imagine how we might adapt the places around us to rising sea levels and temperatures in a world rushing towards resource scarcity. With a little bit of imagination and sufficient resourcing, architecture schools could become nexus points, laboratories, and hubs for weaving together threads of ongoing efforts, upskilling their staff and students in the process – from supporting street-by-street retrofit plans to the reimagining of redundant infrastructure. Setting project briefs that ask any less of students is to deny the reality of where we are.

Climate action in construction is not just about reducing emissions, energy demand, or embodied carbon, or even adding greenery (as it is, far too often, unlikely to survive). It is not about capitalising on PR opportunities afforded by the COP process, greenwashing each Earth Day and continuing as usual for the rest of the year, championing the design of airports, and remaining productively complicit in the demolition of our cities. It is about architects and educators playing our part in saving lives and stewarding what already exists. It is time the industry and its education systems understand and embody this.

What really shapes buildings

Contrary to architectural dogma, its education system, and popular mythology, form has never followed function. Buildings are shaped by access to energy, culture, material availability, and most fundamentally, power. Architecture physically expresses our unsustainable civilisation,[20] with almost all buildings remaining dependent on fossil fuels to some degree and sustainability remaining voluntary.

Buildings are under construction and in development today that will require retrofitting within decades – if they are left standing. Most of these buildings will not be as energy efficient as Passivhaus has proved possible since the 1990s, they will not soften the impacts of rising energy bills as fuelled by profiteering energy companies, they will not be designed to be deconstructable so that their materials would have numerous lives, and they will not be prepared to safeguard us from a warming world. Yet, there is little to no outrage, even less resistance, to this in architecture and construction. In its silence, architectural education remains complicit in the lack of public awareness of the matter.

In their essay "Waste of Space", Caroline O'Donnell and Dillon Pranger extend the invitation that architects can "affect, through the legibility of our work, the behaviours and the policies that shape our futures".[21] Emphasising that the places we build are not only the backdrop to our lives, but these spaces are a key determinant of the possible lives we can live, the impact those lives have environmentally, and even of how long those lives are. We must not forget that how we design and build today plays a part in setting the scene for what becomes possible tomorrow and for decades to come.

Today, 40% of any nation's healthcare costs are attributable to its buildings,[22] and this means that architecture and the stewardship of built spaces play a fundamental role in shaping societal health and well-being. Embracing O'Donnell and Pranger's vision involves shouldering this responsibility for our communities just as much as it unlocks an abundance of opportunity. Fusing compassionate, evidence-informed, and circular design would empower us to use far fewer resources to do far greater good – to lighten ecological footprints and take collective care of one another at a time of cascading crises.

Nevertheless, before design is even on the table, built environments are shaped by an ecology of dark matter. Buildings are shaped by building regulations and how they are enforced;[23] by government policies, which are devised by those with access to power; by economic interests;[24] by ownership of land and other buildings; by power itself; and by the worldview of those wielding it.[25] Beyond this, buildings are also shaped by existing physical infrastructure, over which they might exert influence into the future – like roads, public transportation (or lack thereof), electricity grid capacity, and fossil fuel dependencies.

Unfortunately, in today's economic and political landscape of uncare,[26] the dark matter shaping buildings does not place the well-being nor thriving of life at its heart. Unless this changes, construction as we know it shall continue to do damage beyond comprehension by default. We urgently need education systems to critically engage with this; otherwise, architecture students will remain in the dark about what carrying on with business-as-usual in construction really means, for people and planet.

Agency and entanglement

Addressing architecture's sizeable contribution to climate change and unlocking its wealth of potential for Just Transition opportunities require collective action far beyond the building-by-building scale. To get there, we must embrace our agency to affect change – as architects, activists, or as educators – recognise our entanglements, and step into our own power. Easier said than done, considering that in today's political climate, we live in what Mariame Kaba and Kelly Hayes describe as "a society that has been locked into a false sense of inevitability".[27]

Crucially, our civilisation's infrastructure and systems all began in the imagination – whether they take on digital, physical, or social forms – and are, therefore, inherently malleable. Jayne Engle, Julian Agyeman, and Tanya Chung-Tiam-Fook unpack this principle further in *Sacred Civics*, explaining that "everything besides sacred natural laws is socially constructed, and therefore in the realm of possibility to change".[28] Whether it is a political system, an energy system, a planning policy, a decision on a demolition, a hostile environment, or a pipeline, if it began in the human imagination, the reality is that it can change, and can be changed.

Positively affecting the design of buildings could radically reduce future emissions, and we need to drive them down at pace. Numerically, the average large construction project has an embodied carbon impact 196 times larger than an average U.K. resident's annual carbon footprint.[29] Meaning that, during the average working day, architectural workers have potential to significantly reduce carbon emissions multitudes more so than by making changes to their lifestyle, and students must learn this. We have so much more power than we have been taught thus far. Realising more regenerative futures requires the embrace of personal and collective agency, and of stepping up: studio spaces must become an empowering and safe space to practice this.

Cultivating agency was explored during "Reimagining Futures: Activism, Agency, and Provocation" with a cohort of students at Manchester School of Architecture's "Some Kind of Nature" studio in 2021. The Anthropocene Architecture School was invited to run pre-Week-1 workshops, offering students radically different framings for their subsequent studio projects, involving the integration of nonhuman life.

My own contribution was a lecture punctuated by seminar sessions to set our scene, which explored the climate crisis, occasions when cities were shaped through collective action, and opportunities for intervention in decision-making. After follow-up seminars – during which I encouraged students to take their research onto the streets of Manchester – the team's contribution to the Studio Atlas,[30] a document that pooled the outputs from each guest workshop, was beautifully illustrated evidence of each student feeling significantly more empowered (Figures 1.1 and 1.2). Without agency, learners wait for permission to apply and exercise knowledge: today, we do not have the luxury of waiting for that permission.

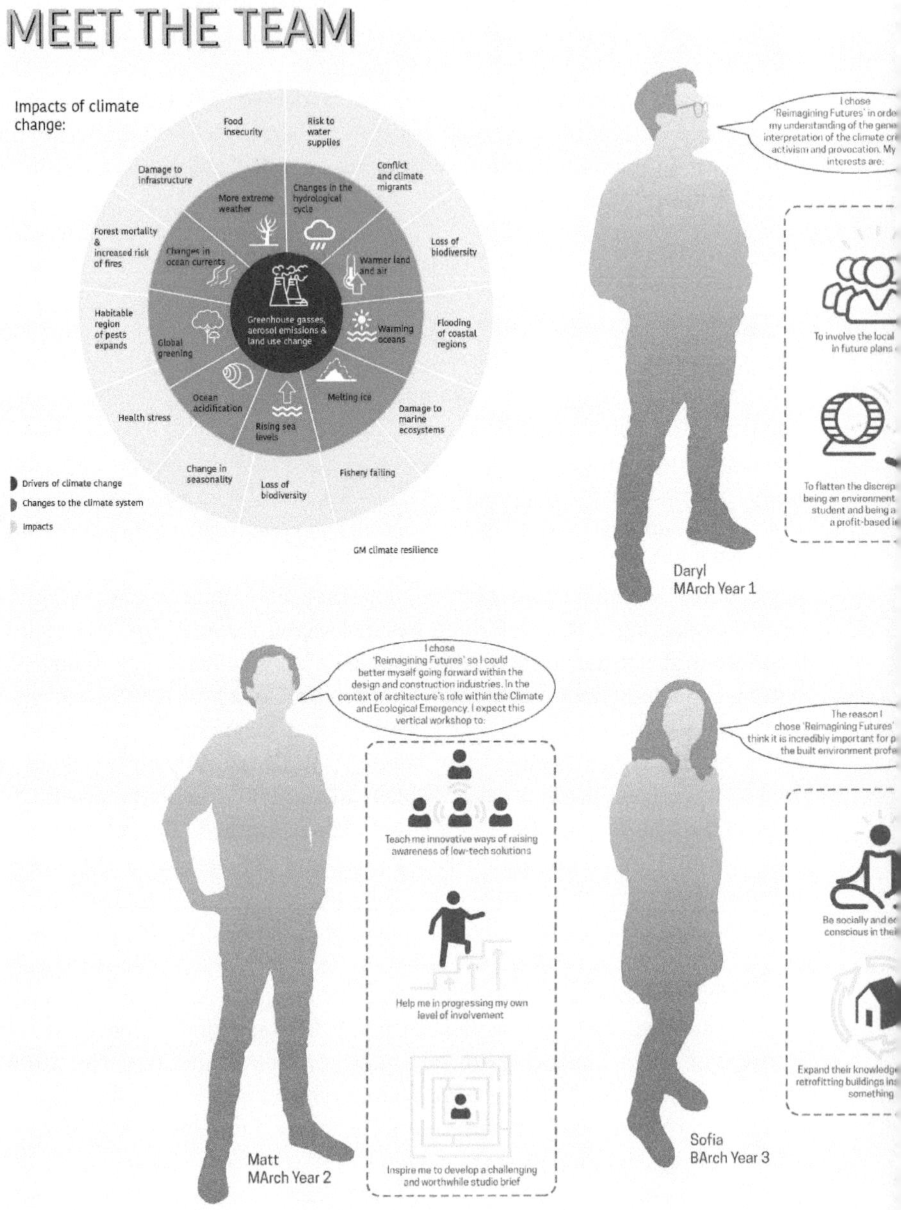

FIGURE 1.1 "Reimagining Futures" student output before "Some Kind of Nature" 2021–2022

Source: Abdullah Alamoudi, Mohamed Ahmed Harfoush, Hrithik Aggarwal, Matthew J Crossley, Daryl Quayle, Claudia Rowe, and Sofia Viudez Solé

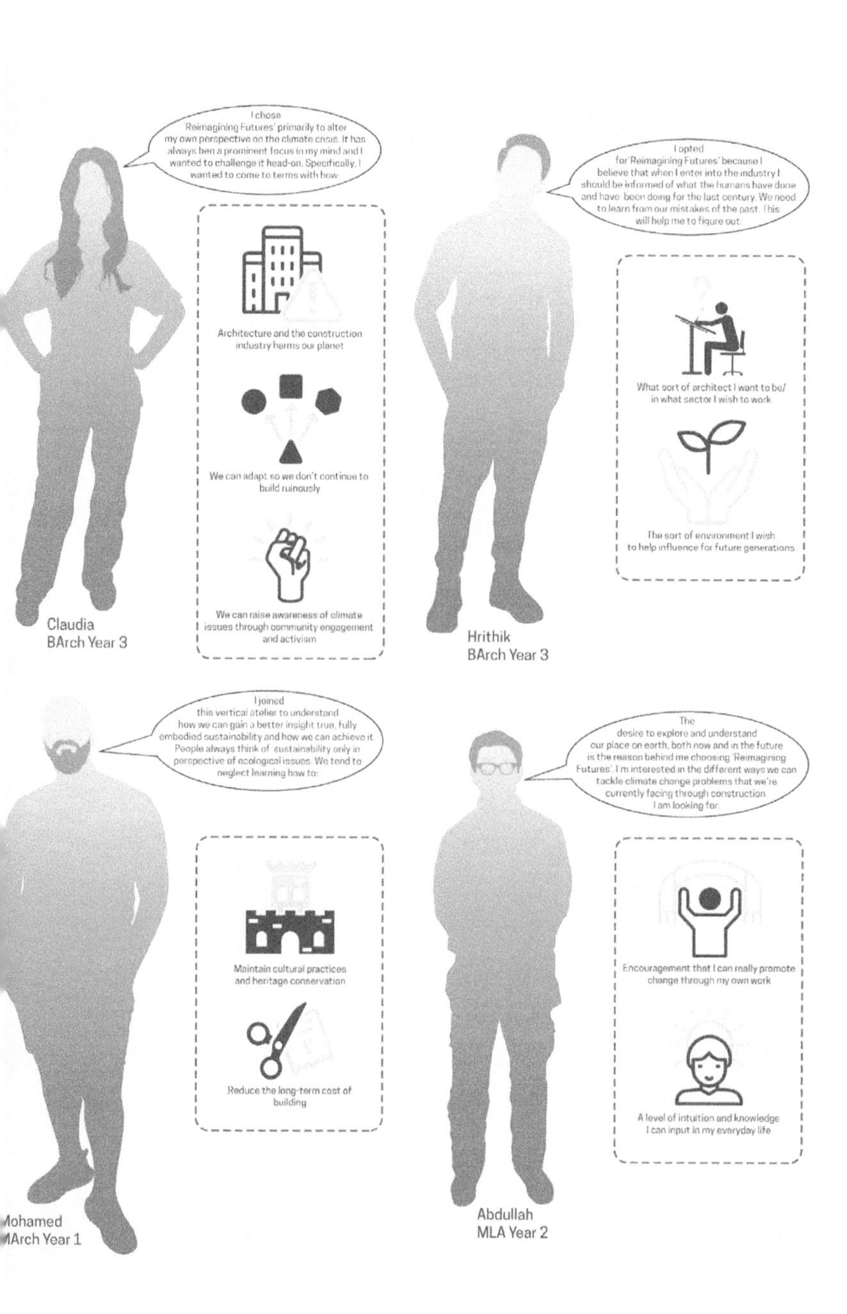
I chose 'Reimagining Futures' primarily to alter my own perspective on the climate crisis. It has always ben a prominent focus in my mind and I wanted to challenge it head-on. Specifically, I wanted to come to terms with how:
Architecture and the construction industry harms our planet
We can adapt so we don't continue to build ruinously
We can raise awareness of climate issues through community engagement and activism
Claudia
BArch Year 3
I opted for 'Reimagining Futures' because I believe that when I enter into the industry I should be informed of what the humans have done and have been doing for the last century. We need to learn from our mistakes of the past. This will help me to figure out:
What sort of architect I want to be/ in what sector I wish to work
The sort of environment I wish to help influence for future generations
Hrithik
BArch Year 3
I joined this vertical atelier to understand how we can gain a better insight true, fully embodied sustainability and how we can achieve it. People always think of sustainability only in perspective of ecological issues. We tend to neglect learning how to:
Maintain cultural practices and heritage conservation
Reduce the long-term cost of building
Iohamed
IArch Year 1
The desire to explore and understand our place on earth, both now and in the future is the reason behind me choosing 'Reimagining Futures'. I'm interested in the different ways we can tackle climate change problems that we're currently facing through construction. I am looking for:
Encouragement that I can really promote change through my own work
A level of intuition and knowledge I can input in my everyday life
Abdullah
MLA Year 2

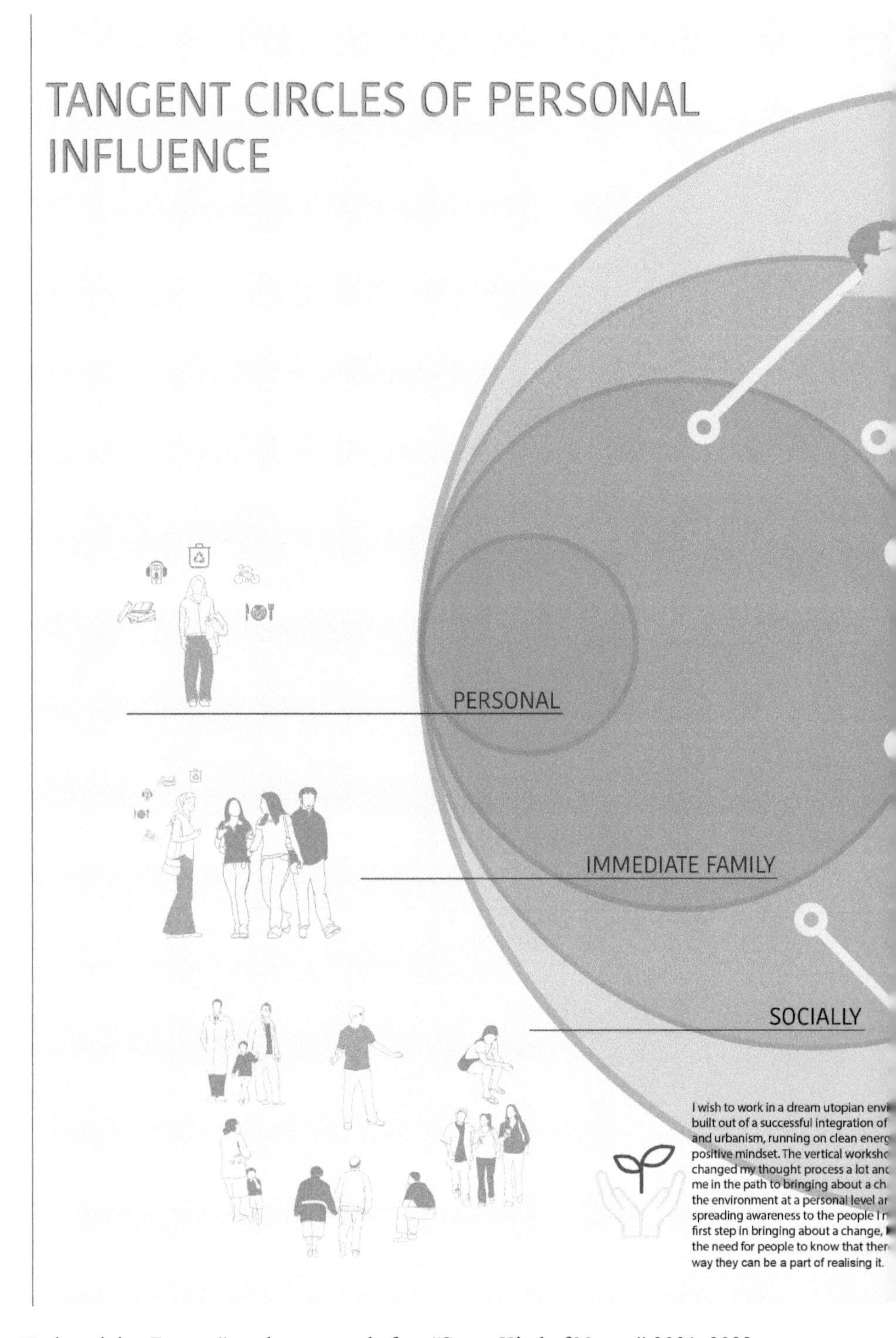
TANGENT CIRCLES OF PERSONAL INFLUENCE
PERSONAL
IMMEDIATE FAMILY
SOCIALLY
I wish to work in a dream utopian
built out of a successful integration of
and urbanism, running on clean
positive mindset. The vertical
changed my thought process a lot
me in the path to bringing about a
the environment at a personal level
spreading awareness to the people I
first step in bringing about a change,
the need for people to know that
way they can be a part of realising it.

FIGURE 1.2 "Reimagining Futures" student output before "Some Kind of Nature" 2021–2022

Source: Abdullah Alamoudi, Mohamed Ahmed Harfoush, Hrithik Aggarwal, Matthew J Crossley, Daryl Quayle, Claudia Rowe, and Sofia Viudez Solé

has always been the main driver for each project
ourney in the MSA. It wasn't until I was shown the
that I realised how inactive I have been with my
ng Futures workshop has helped me realise the
mmunity engagement within larger urban projects,
influence and my future projects both in university
ctice.

I have always believed that climate change is a ticking bomb that we all need to deal with. When I was first exposed to the diagram, I was in the zone where no change can really happen unless there is a more political will. After this workshop I realised that we are capable of addressing the issues we face through a more proactive communication and more involvement of communities.

I've always been aware of the effect that climate change has on the world but I wanted to understand and learn more. This workshop enlightened me on how important it is to communicate and engage with the community in order to make a change. It gave me the opportunity to be more proactive and knowledgeable in this topic which will help me create more eco-friendly and sustainable designs.

This workshop inspired me to go out and talk to people about construction and climate change. By involving the community, we can build where it matters most and prevent unnecessary construction. Through this workshop I was able to understand the concerns and fears I have around climate change and re-evaluate my perspective to have a more optimistic approach. I have also been inspired to protest and take an active stance in raising awareness of the climate crisis.

What I have gotten out of this workshop is that I should strive to interact with the city and citizens more within my designs going forward as at the end of the day our designs are meant for the public. Going forward I will concentrate on implementing reftrofitting and reusing infrastructure in my designs as I think it is incredibly important that this be a key element in my designs, while still being ecologically and socially conscience.

This workshop has given me the confidence to challenge myself in the up-and-coming studio work. I really want to push for, not just an environmental carbon-neutral solution, but true 'climate positive' architecture that features low-tech components. This shall be a difficult task to undertake, yet it shall also be exciting. I now believe it is essential if I am to be venturing back into practice within the next 12 months, I must pay close attention to Scott's teachings.

ISM

Teaching students where architecture and construction intersect with the climate crisis must not be done in isolation from the systems in which any architect is entangled, culturally, socially, economically, and politically. It cannot be reduced to a technical issue or kept out of the design studio, and this knowledge must develop in parallel with exploring how overlapping social, economic, and political systems affect architectural practice, how systems of oppression manifest, and where we can consciously start to break these cycles. There will be little possibility of them implementing that knowledge equitably otherwise. Such work requires space and time for articulation, for unlearning what no longer serves us, and importantly, for imagining the future otherwise.

Cultivating the radical imagination

Despite sufficient technology being available to build low-carbon futures where we could take better care of one another and the world around us, invitations or moments to imagine them remain scarce, especially within architectural education. Rob Hopkins, a founder of the Transition Network, suggests that "we need to be able to imagine possible, feasible, delightful versions of the future before we can create them. Not utopias, but where things turned out okay".[31] If embraced, such framing could prove transformative in design studios and the world beyond them.

In *Crises of Imagination, Crises of Power*, Max Haiven defines the imagination as "a shared landscape or commons of possibility", and, whilst the practice of the radical imagination can take many forms, at its core, it is "a matter of acting otherwise together".[32] Such collaborative thinking – an embracing of disaster collectivism – offers us the best chance at responding to the climate crisis. This demands the antithesis of the individualistic, heroic genius that architecture and its education have celebrated and romanticised for centuries. To work as collaboratively as is required will need an unlearning of such individualism.

We know that we can realise healthy, low-carbon, and enchanting buildings and retrofit existing homes to have almost negligible heating bills. These are not utopian dreams; they are today's technologically possible realities. That these are not legal bare minimums is an outrage, but we must remember that many people – architects and many architectural educators included – are not fully aware of what we are capable of when it comes to buildings because they have no lived experience of it, and construction literacies are rarely shared.

In *To Build a Beautiful World, You First Have to Imagine It*, Mary Annaïse Heglar invites readers to practice world-building – a way in which the radical imagination could be fused powerfully with architectural education and practice. She distills this as the practice of enabling and inviting people to experience the world as you imagine it could be; world-building is "as much about creating new things as it is about destroying old structures and assumptions".[33] Education must begin synthesising such lessons so that students and certified architects have opportunities to imagine futures in which sustainable design is the norm rather than an anomaly, in which communities have embraced deep social change – akin to Ernest Callenbach's visionary novel *Ecotopia* – and in which we come together and things do turn out okay.

Exercises in the radical imagination are incredibly powerful and should be woven into architectural education from its beginning. One that can be brought into a lecture – inspired by a time-travelling exercise from podcast *From What If to What Next* – has three components:

1. Bring examples of architectural climate solutions into narrative-led lectures, emphasising stories of the possible, rather than graphs and climate impact photographs.
2. Design in a 5–10 minute-long conclusion, in which you extend an invitation to your students to imagine a day in 2030 if all these climate solutions have become entirely commonplace.

Encourage them to pay close attention to how changes to buildings would affect them as they go about their days.

3. Afterwards, invite them to share their imagining with another student, and play it by the room's energy – momentarily losing control of the room to conversations at this point is not a bad sign.

Other ways to nurture your students' radical imagination include:

- being upfront about your own knowledge gaps – what you do not know and learn alongside them by attending sessions that others hold and otherwise
- encouraging and enabling collaborative projects between students
- facilitating seminars exploring sci-fi and speculative fiction
- resourcing cohorts to invite guest lecturers with the financial support to pay those guests
- reimagining design briefs as the answering of questions that meet real community needs
- actively supporting disruption, divergence, and getting involved in real-world campaigns and struggles

Despite quite literally being tasked with designing the future, architecture students are rarely – if ever – asked, or given time, to imagine what it might look, feel, smell, sound, or taste like. Limiting global warming requires "rapid, far-reaching and unprecedented changes in all aspects of society",[34] and that means bold, imaginative actions, not the incrementalism offered by most governments. Changing this radically requires the challenging, unpacking, and unlearning of the social conditioning and mythologies that are holding us back, almost as much as acquiring new skills.

Unlearning powerlessness

Readying architecture students to design and realise sustainable buildings amidst an unpredictable future is not solely a case of bolting additional modules onto today's education system. It necessitates reimagining how architectural culture, design, and history are taught; questioning who and what is celebrated and why; recognising who has been overlooked; leaving toxic cultures in the past; and repairing the harm they have done. But beyond culture and curriculum change, schools of architecture must begin to support students in unlearning the mythology of their own powerlessness; otherwise, they will struggle to step into their power and take any action at all.

Doing this requires nothing short of a cultural transformation in architecture schools and the wider industry beyond, as advocated for by ACAN and Architects Declare. When invited by MOULD to undertake an "Architecture Is Climate" residency at Central St. Martins, I drew together threads of what un/learning towards this end could look like – involving elements of participation, play, and provocation. The Unlearning Powerlessness[35] programme involved Nick Newman from U-Build with whom students and contributors alike built the exhibition infrastructure; Simeon Shtebunaev of Urban Imaginarium facilitating a large-scale play-along of the retrofit board game Climania[36] (Figure 1.3); workshops co-designed by the Doughnut Economics Action Lab facilitated by ACAN and Architects Declare; a lecture from Kelly Doran on the Ha/f Studio[37] – an embodied carbon exercise already affecting policy change in Toronto; and Anthropocene Architecture School workshops. The programme ended with a panel discussion involving architects, activists, engineers, publishers, and trade unionists that raised the bar for all future architecture and climate panels (Figure 1.4).[38] Every aspect of this programme was immediately replicable and could be requested and actioned by any institution without waiting on full curriculum change.

FIGURE 1.3 “Climate education through play” led by Simeon Shtebunaev (Urban Imaginarium)

Source: Simeon Shtebunaev

FIGURE 1.4 Unlearning Powerlessness closing panel on 15 March 2023 with Phoebe Plummer (Just Stop Oil), Morgan Trowland (Just Stop Oil) – not visible, joining by phone from HMP Chelmsford, Aska Welford (UVW-SAW), Martha Dillon (Positive Money), and Tom Bennett (Architects Climate Action Network, Project Bunny Rabbit, and Studio Bark), chaired by Scott McAulay (Anthropocene Architecture School).

Source: James McAulay

Coming full circle

Now, imagine yourself as an architecture student today: working towards a degree, fully aware of climate breakdown taking place, experiencing evidence of it regularly, knowing the industry does huge environmental harm, and your course is not offering you actionable solutions. The most powerful analogy I have heard has given this chapter its name: a Glaswegian student, Carmen Lean, remarked that "it's like we're being taught to design our own coffins". It is no wonder, faced with such experiences, that this student, and many others, are gravitating towards community work or direct-action campaigns rather than completing degrees preparing them for worlds that no longer exist.

We will know that architectural education is moving in the right direction when students leave it feeling as if they have the capacity to radically improve the world around them, despite the house being on fire. That is when we will know that we are teaching students the right stuff. Until then, there is much work to do, that must be done far more collectively, compassionately, and imaginatively than it has ever been done before.

Notes

1 Lizzie Crook. "Students 'Let Down by Their Architectural Education' Says Climate Action Group." *De Zeen*, 2021. www.dezeen.com/2021/03/10/stucan-launch-architects-climate-action-network/

2 Richard Waite. "'Students Told to Ignore Existing Building' – Survey Reveals Retrofit Teaching Gap." *Architects' Journal*, 2022. www.architectsjournal.co.uk/news/students-told-to-ignore-existing-building-survey-reveals-retrofit-teaching-gaps

3 World Green Building Council (WGBC). "Bringing Embodied Carbon Upfront." 2019. https://worldgbc.org/advancing-net-zero/embodied-carbon/

4 Architects Climate Action Network (ACAN). "Climate Curriculum Campaign." 2023. www.architectscan.org/curriculum-campaign

5 Naomi Klein. *On Fire: The Burning Case for a Green New Deal*. London: Penguin Audio: Audible, 2019.

6 Simon Willcock, Gregory S. Cooper, John Addy and John A. Dearing. "Earlier Collapse of Anthropocene Ecosystems Driven by Multiple Faster and Noisier Drivers." *Nature*, 2023. www.nature.com/articles/s41893-023-01157-x

7 InfluenceMap. "Big Oil's Real Agenda on Climate Change 2022." 2022. https://influencemap.org/report/Big-Oil-s-Agenda-on-Climate-Change-2022-19585

8 Dharna Noor. "Big Oil Quietly Walks Back on Climate Pledges as Global Heat Records Tumble." *The Guardian*, 2023. www.theguardian.com/us-news/2023/jul/16/big-oil-climate-pledges-extreme-heat-fossil-fuel

9 Intergovernmental Panel on Climate Change (IPCC). "AR6 Synthesis Report (SRY)." 2023. www.ipcc.ch/report/sixth-assessment-report-cycle/

10 Centre for Alternative Technology (CAT). "Zero Carbon Britain: Rising to the Climate Emergency." 2019. https://cat.org.uk/info-resources/zero-carbon-britain/research-reports/zero-carbon-britain-rising-to-the-climate-emergency/

11 United Nations (UN). "Speech: Mia Mottley, Prime Minister of Barbados at the Opening of the #COP26 World Leaders Summit." 2021. www.youtube.com/watch?v=PN6THYZ4ngM

12 Climate Action Tracker. "The CAT Thermometer." 2022. https://climateactiontracker.org/global/cat-thermometer/

13 Oxfam. "Billionaires Responsible for a Million Times More Greenhouse Gases than the Average Person." 2022. www.oxfam.org.uk/mc/9b2op5/

14 Duncan Baker-Brown. *The Re-Use Atlas: A Designer's Guide Towards a Circular Economy*. London: RIBA Publishing, 2017. pp. 7–15.

15 Daniel Christian Wahl. *Designing Regenerative Cultures*. Dorset: Triarchy, 2016. pp. 24–25.

16 Jason Hickel. *Less Is More: How Degrowth Will Save the World*. London: William Heinemann, 2020. pp. 78–79.

17 Julia Watson. *Lo-TEK: Design by Radical Indiginism*. London/Los Angeles: Taschen, 2021. pp. 20–27.
18 Amanda Sturgeon. "Buildings Designed for Life." In *All We Can Save: Truth, Courage, and Solutions for the Climate Crisis*, edited by Ayana Elizabeth Johnson and Katherine K. Wilson. New York: Penguin Random House, 2020. p. 166.
19 Susannah Hagan. *Revolution? Architecture and the Anthropocene*. London: Lund Humphries, 2022. p. 93.
20 Barnabus Calder. *Architecture: From Prehistory to Climate Emergency*. London: Pelican Books, 2021. p. 432.
21 Caroline O'Donnell and Dillon Pranger. "Waste of Space." In *The Architecture of Waste: Design for a Circular Economy*, edited by Caroline O'Donnell and Dillon Pranger. Oxford: Routledge, 2020. p. 43.
22 Martin Brown. *Futurestorative: Working Towards a New Sustainability*. London: RIBA, 2016. p. 16.
23 Peter Apps. *Show Me the Bodies: How We Let Grenfell Happen*. London: Oneworld, 2022. pp. 340–341.
24 David Madden and Peter Marcuse. *In Defense of Housing: The Politics of Crisis*. New York and London: Verso, 2016. Kindle. p. 9.
25 Daniel Christian Wahl. *Designing Regenerative Cultures*. Dorset: Triarchy, 2016. pp. 131–132.
26 Sally Weintrobe. *Psychological Roots of the Climate Crisis: Neoliberal Exceptionalism and the Culture of Uncare*. New York: Bloomsbury Academic, 2021. pp. 111–116.
27 Mariame Kaba and Kelly Hayes. "A Jailbreak of the Imagination: Seeing Prisons for What They Are and Demanding Transformation." *truthout*, 2018. https://truthout.org/articles/a-jailbreak-of-the-imagination-seeing-prisons-for-what-they-are-and-demanding-transformation/
28 Jayne Engle, Julian Agyeman and Tanya Chung-Tiam-Fook. "Imagine Shaping Cities as If People, Land, and Nature Were Sacred." In *Sacred Civics: Building Seven Generation Cities*, edited by Jayne Engle, Julian Agyeman and Tanya Chung-Tiam-Fook. Abingdon: Routledge, 2022. p. 7.
29 Architects Declare. "Architects Declare Practice Guide Volume 1." 2021. www.architectsdeclare.com/uploads/AD-Practice-Guide-2021-v1_3.pdf
30 Atelier Some Kind of Nature. *Some Kind of Nature Atlas*. Stockport, Manchester: Manchester School of Architecture, 2021 (unpublished).
31 Rob Hopkins. *From What Is to What If: Unleashing the Power of Imagination to Create the Future We Want*. Hartford, VT: Audible, 2020.
32 Max Haiven. *Crises of Imagination, Crises of Power: Capitalism, Creativity and the Commons*. London and New York: Halifax and Winnipeg, Fernwood Publishing, Zed Books, 2014. p. 218.
33 Mary Annaïse Heglar. "To Build a Beautiful World, You First Have to Imagine It." *The Nation*, 2022. www.thenation.com/article/environment/climate-world-building/
34 Intergovernmental Panel on Climate Change. "Special Report Global Warming of 1.5°C." 2018. www.ipcc.ch/sr15/
35 Mould. "Anthropocene Architecture School – Unlearning Powerlessness." 2023. http://v2.mould.earth/aic-exhibition-aas
36 "Climania: The Climate Action Board Game." https://climaniathegame.com/
37 Kelly Doran. "A Whole Life Approach." *Architecture Is Climate* (lecture, University of Central Saint Martins, London, 15 March 2023). https://youtu.be/9Me-OLW5xbg
38 Anthropocene Architecture School. "Unlearning Powerlessness." *Architecture Is Climate* (panel discussion, University of Central Saint Martins, London, 15 March 2023). https://youtu.be/Qx6l1gaMA4A

2

RE-USE AESTHETICS AND THE ARCHITECTURAL ROOTS OF ECOLOGICAL CRISIS

Ben Sweeting

Aesthetics of re-use

Architectural design is complicit in the planet's enveloping ecological crisis. The energy embodied in and consumed by buildings is a major contributor to carbon dioxide emissions, driving climate change. Buildings destroy habitats through the land they occupy, the pollution they cause, and the resources they extract, leading to biodiversity loss. Attempts to mitigate these harms have been primarily framed through a technical discourse concerned with efficiencies of matter and energy, and understandably so. However, to focus only on matter and energy is to accept, at best, a limited understanding of ecology and its relationship with architecture. Given the prominence of this technical framing, the way that aesthetic considerations feature in the work collected in this volume is notable. A circular economy for the building industry could be implemented without making a difference to how buildings look and feel. From the technical framing of matter and energy, what buildings look like is of minor importance. Indeed, aesthetic considerations of any kind might be dismissed as a distraction in the context of a contemporary crisis that demands immediate practical action. But, for many of the projects presented here, it matters not just that material is re-used but that this re-use is experienceable. Is this "just" an aesthetic?

In a sense, the emerging re-use aesthetics is a kind of style – one that returns to the modernist tradition of developing architectural interest out of technology but now reflecting a radically different mode of construction. The re-use of what would otherwise become waste as construction material offers many aesthetic possibilities, from ad hoc juxtapositions to the embodiment of cultural memory. These qualities are valuable in their own terms, independent of the practical benefits of re-use. Are there also ways that the aesthetics of re-use might contribute to the environmental agenda that they reflect? One way of thinking about this possibility is in terms of promoting the wider adoption of re-use. If material circularities are valuable for more than just practical reasons, then this, in turn, makes re-use more feasible and desirable.[1] In this way, the aesthetic qualities of re-use can support its adoption, such that attention to aesthetics pays off even in technical terms. From this perspective, the environmental value of re-use aesthetics is

DOI: 10.4324/9781032665559-6
This chapter has been made available under a CC-BY-NC-ND 4.0 license.

as a matter of advocacy – something to be judged, in the end, through the logics of matter and energy. There is, I suggest, more to it than this.

Rather than thinking of re-use in terms of style, the focus on the aesthetic that I develop in this essay is rooted in a concern with systems that is implicit in the idea of the circular economy (and sustainability transitions more generally). Systemic and aesthetic considerations are not usually associated with each other, so, perhaps, this will seem odd at first. Creating circular economies of materials entails shifting focus from the design of individual buildings to the design of the building industry, and so to the political, social, economic, infrastructural, and environmental systems in which the building industry is embedded. Decentring individual building projects in this way would seem to also decentre the aesthetics of those projects. But in another sense, the aesthetic qualities of architecture are full of possibilities for systemic change – for transforming the understandings, assumptions, and relationships within which architecture and much else is made. What I outline in this essay is an expanded reading of the ecological relevance of architecture, where the aesthetic qualities of buildings are understood to make a difference in their own terms, for good or ill; something to be considered alongside technical concerns with matter and energy.

In making this argument, I draw on the work of anthropologist and cybernetician Gregory Bateson (1904–1980), whose later work was developed in the context of the emerging environmental consciousness of the 1960s and 1970s. Bateson was a key figure in the development of cybernetics, an unusual transdisciplinary field that contributed to the foundations of contemporary ways of understanding systems, including in design.[2] Both cybernetics and re-use are concerned with circularity. Whereas re-use is concerned with material, cybernetics is focussed on informational and organisational circularities across multiple domains, including ecological, biological, social, and technological contexts. An example of cybernetic circularity is feedback, where the outcomes of action are taken as inputs for further action, forming a causal loop. Feedback can lead to the maintenance of conditions in a changing environment, such as steering a steady course when sailing a ship. Feedback can also lead to runaway, such as increasing global temperatures leading to the melting of polar ice caps, which, in turn, leads to less sunlight being reflected away, further increases in global temperatures, and so further melting.

Bateson understood one of the root causes of ecological crisis as the "hubris" of Western culture's tendency to see humans as separate to, above, and in competition with their environment and each other, contrasting this attitude with indigenous ways of knowing.[3] While environmental challenges have shifted focus and intensified in the decades since Bateson was writing, this recognition of hubris as an underlying cause remains pertinent. Hubris can be reinforced even in well-meaning attempts to be sustainable, with the result that it is hard to identify and address. The aesthetics of re-use have a potential role to play in challenging hubris, complementing the immediate practical contributions of circular economies in mitigating the harms caused by the built environment.

Bateson understood ecology in terms of patterns of learning and communication through which organisms and environments are interrelated. The difficulty of appreciating these patterns within a conventional scientific focus on matter and energy led to Bateson arguing for the need to incorporate aesthetic concerns in the developing science of ecology.[4] In this sense, aesthetics is not style or appearance, or a matter reserved just for art critics, philosophers, or even humans,[5] but, instead, a sensitivity to and empathy with the patterns that connect all living things.[6]

While Bateson was focussed on the context of science, similar concerns are also relevant in designing architecture. It is not just science that shapes humans' understandings of the worlds they inhabit. The built environment does this, too, through the ways in which built spaces frame everyday experiences. In so doing, the built environment can both obfuscate and uncover ecological patterning. It is through this lens that I propose understanding the aesthetic qualities of re-use in architecture. Not as style or advocacy. But as a way in which aspects of ecological pattern can be made tangible within everyday experiences.

Roots of ecological crisis

As a point of departure, I take Bateson's "The Roots of Ecological Crisis", which originated as testimony to a committee of the State Senate of Hawaii in 1970 and was subsequently published in *Steps to an Ecology of Mind*. In a style that is more direct than much of Bateson's other work, this short text warns against ad hoc solutions to ecological problems that focus on symptoms and leave underlying causes in place. Bateson identifies three underlying root causes of ecological crisis, naming these as population, technology, and hubris.[7] These each have the potential to be self-reinforcing and reinforcing of each other, producing runaway feedback.

The first of these three roots, population, needs some reframing from the context of the 1970s in which Bateson was writing. Bateson refers to "population increase" and "the population explosion", which were concerns of the time.[8] It is not the number of humans per se, however, but the demands that the human population places on the planet through consumption that are at issue. This root of ecological crisis is thus best understood as growth in a general sense.[9] The demands humans place on the planet are globally and socially unequal. They can accelerate even for a static population because of commitments to economic growth and rising living standards. While one can conceive of the rate of increase of these demands slowing (degrowth, more efficient use of resources, increasing environmental awareness), it is difficult to imagine how to halt or reverse this growth in a managed way. Many claims to reduce carbon emissions are actually claims to increase them by less than one would have done otherwise. It is politically difficult to agree on courses of action such as reducing living standards, and there are ethical difficulties in doing so because of the intersection with social and global injustices.

The second root of ecological crisis that Bateson identifies, technology, is closely linked to growth. The demands of human society drive developments in technology. In turn, technologies make other technologies possible, permitting and prompting the demands of human society to grow further. The rapid technological developments of recent centuries have led to widespread pollution, ranging from microplastics to carbon dioxide emissions.

In one sense, of course, technological developments can be a way to reduce demands that humans place on their environment, for instance, by making energy usage more efficient or less polluting. But because of the reinforcing feedback between technology and growth, technology will always have a double relation to ecological crisis. Like growth, it is difficult to imagine how to undo technological change, as humans become dependent on the technologies they develop. For instance, it is not possible to simply step back to pre-industrial agriculture because society has become dependent on the increased yields that industrialisation made possible. Similarly, in sustaining itself, the design industry is structurally committed to innovation, profit, and narratives of technological progress. Bateson articulates ad hoc technological measures as a kind of addiction.[10]

The third root of ecological crisis that Bateson identifies is the hubris of Western culture's conception of humans as separate to and in competition with the environment and each other. Hubris, excessive pride or overconfidence, is less tangible than growth or technology. It is one part of the peripeteian structure (the dramatic reversal of fortune) of Greek tragedy, which is apt as an analogy for human-made ecological catastrophe. Hubris is made manifest in attempts to bend ecosystems to human will. The most obvious examples are where humans destroy their environments in extracting resources or making human habitats. Hubris is closely coupled with both growth and technology. Hubris motivates control over ecosystems through technology and supports the extension of human activities at the expense of the wider ecosystem. It is reinforced by "success" in these endeavours. Sustainable design itself is not immune from hubris. Even well-meaning attempts to address ecological problems can be manifestations of hubris in the sense that they can proceed from humans' faith in their own expertise and capacity for unilateral action.

The most extreme example of hubris is geoengineering – the use of technology to control the temperature of the planet in response to the climate crisis. One scenario is the continual injection of sulphate aerosols into the stratosphere, providing a cooling effect by preventing sunlight from reaching earth. As the climate crisis worsens, well-meaning proposals for geoengineering projects such as this will gain traction as technological responses to rising temperatures. The limitations of geoengineering proposals are most obviously their treatment of symptoms (temperature) rather than causes (greenhouse gas emissions). The justifications given for geoengineering recognise this limitation, arguing that such projects buy time while other technological solutions are developed. However, it would be difficult to stop geoengineering once it has been started, without creating sudden shocks.

By focussing on temperature, geoengineering mischaracterises the issue it seeks to address in more ways than one. While greenhouse gas emissions and the failure of the world to reduce these may be causes of rising global temperatures, they are also *symptoms* of deeper problems. The problem of geoengineering is not just that it treats only symptoms but that it *intensifies the underlying causes* of crisis by reinforcing hubris.

Geoengineering projects would completely enframe the planet using technology. One might even say that geoengineering would effectively turn the planet into a building – an interior space – subject to human will at planetary scale. This intensification of causes is the case even if geoengineering was "successful" in its own terms, as this would build further human overconfidence in being able to manipulate ecosystems through technological expertise. How would subsequent manifestations of ecological crisis be responded to? The consequences of geoengineering are not just for climate but for how humans understand their relation to the ecologies of which they are part. It is possible to save the planet in ways that go on to destroy the planet, depending on *how* one does the former. Actions are not just what we do but the things that are produced by the ways in which we act, even when these are unintended. Means are also ends, having consequences.

Geoengineering would have uneven global effects. Reducing the amount of sunlight falling on the planet would have variable consequences in different parts of the globe. Yet the power to decide over how to do this would rest with those countries that control the technological infrastructure. If enacted, geoengineering would be one more instance of the way that the hubris of Western culture has supported and been propagated by processes of marginalisation and colonialism.[11]

Hubris is deeply rooted in Western culture and especially in its epistemology – its ways of thinking and knowing. It is not that conventional Western epistemology is the only shit epistemology. But it is the shit epistemology that has been globalised through colonialism and the internationalisation of design during the twentieth century. A limitation of this essay, and my work more generally, is its emplacement within Western culture. I lean here not just on a Western thinker (Bateson) but also on examples from the Eurocentric architectural traditions within which I have been educated. Yet, unmaking hubris is not as straightforward as adopting an alternate epistemology. To pick up another epistemology without critically addressing one's own is likely to distort the former through the latter. Appropriation is one manifestation of Western hubris, after all.

Bateson suggests that hubris is the easiest of these three roots to reverse and that acting on one of the three may be enough to avert catastrophe.[12] From today's perspective, however, it would seem that hubris is no less straightforward to counter than the other two roots and that it is the feedback relations between all three that need to be addressed. Many sustainable design projects are focussed on technology, growth, and the relation between the two, where efficiencies in technology reduce, mitigate, or at least slow the increase of demands that humans place on the planet. Further attention needs to be paid to hubris and its relations to growth and technology – the ways that they reinforce and are reinforced by hubris. For the most part, mainstream design still operates from within a hubristic conception of growth (the idea that human activity can continue to expand unchecked) and through a hubristic mode of technology (unilateral, instrumental control of ecosystems).

There is a paradox of sorts to hubris. Escaping hubris is not as simple as reaching some "correct" epistemology (as Bateson sometimes seems to imply). To see oneself as having done this would be hubris itself. It is tempting to associate hubris with other people: with adherents to whichever political and philosophical commitments one objects to. Systems thinkers might locate hubris in reductivism, constructivists in realism, progressives in conservatism, and so on. But setting oneself apart from others *is* part of hubris. Thinking one has overcome hubris *is* a form of hubris. To take hubris seriously is to examine one's own part in it. Hubris can even arise in attempts to escape it – the hubris of overcoming hubris, a conflict across what Bateson would refer to as logical types. Given this, it is difficult to address hubris directly.

Aesthetic engagements with ecosystems offer one possible way to bypass this paradox. If hubris is seeing oneself as separate to and in competition with the environment, then hubris can potentially be countered by cultivating perceptions of one's embeddedness within and dependence on the ecosystemic relations of which one is part. Architectural design has something to contribute to such an effort, given its traditional roles in orienting humans within the worlds they inhabit. How might the aesthetics of re-use contribute to this?

Architecture and hubris

The built environment is closely related to all three of the roots of ecological crisis that Bateson identifies. This is most obvious in the cases of the first two, the growth of the demands that humans place on their environments and the development of the technologies through which they do it. Buildings are, after all, one variety of technology. Like other technologies, they consume energy and material resources while producing waste. Buildings are constructed in response to various kinds of growth and, in turn, support and prompt further growth. As well as meeting humans' spatial needs and desires, increasingly buildings even play a role as financial assets, driven by the growth of investments.[13]

Buildings are also closely related to hubris. The built environment prioritises human activities and extends human control. The hubris of the built environment is clearest in large-scale attempts to control climatic conditions through technology: proposals for a geodesic dome over

Manhattan, contemporary plans for building mega cities across deserts, and ultimately, the enframing of the planet through geoengineering.

The built environment can also imply hubris in how it is experienced in a more everyday sense. Hubris is implicitly reinforced by the way that the conventional built environment separates human and ecological worlds. Buildings (literally) construct relations between human and ecological worlds. For the most part, these relations take the form of sharp distinctions (such as walls and windows) that separate out the two, allowing humans to exert control of internal conditions such as temperature as well as who and what enters and leaves. Consider the rooms that you have been inside today as examples of this. Who and what would and would not be welcomed inside? The separating out of human and ecological worlds reinforces hubris by presenting ecological relationships as if they are distinct from the human world.

This separation is sometimes heightened by the aesthetics of architecture. Consider, for instance, the Farnsworth House designed by Ludwig Mies van der Rohe for Edith Farnsworth (Plano, IL, USA; constructed 1945–1951). Its form and material contrast with its surroundings, with the built form held elegantly apart from the ground. The architecture implies separate "human" and "natural" worlds seemingly corresponding to inside and outside respectively (Figure 2.1).

The glass walls open the interior to its surroundings, which Mies expressed in terms of letting the outside in.[14] This mode of apparently connecting to nature actually implies a separation: "Nature, too, shall have its own life . . . If you view nature through the glass walls of the

FIGURE 2.1 The Farnsworth House

Source: Carol M. Highsmith Photographs in the Carol M. Highsmith Archive, Library of Congress, Prints and Photographs Division. LC-DIG-highsm-13100

FIGURE 2.2 The Farnsworth House flooded by the nearby river

Source: Mills Baker. CC BY 2.0. www.flickr.com/photos/millsbaker/2861738008/

Farnsworth House, it gains a more profound significance than if viewed from outside".[15] The implied idea of separation is an erroneous one. What is outside is not natural, and the human world inside is subject to its environment, as becomes explicit when the building is flooded by the nearby river (Figure 2.2).

Bateson expresses hubris not only in terms of humans seeing themselves as against nature but also as against each other. And, indeed, the human-nature hubris performed by the Farnsworth House intersects with human-human hubris in the architect's attitude to the client. Mies treated Farnsworth as "a means to an architectural end",[16] with the architecture that sets up the binary contrast with nature also neglecting the client's needs and desires in its absence of programmatic consideration.

A more complex example is the Flower Tower, an apartment block designed by Edouard Francois (Paris, France; completed 2004, Figure 2.3). This building might be casually called "green" because of its use of planting as part of its facade. It is claimed that this "embodies the expression of desire for nature in the city",[17] and it succeeds in blurring the boundaries between architecture and environment. It seems like this is the sort of architecture which makes the relation between humans and nature more ambiguous, an antidote to the sharp boundaries of the Farnsworth House.

But what idea of nature is implied by this building? What idea of humans' place in the world is being performed? A species from somewhere else on the planet (bamboo) is enframed within the architecture, with giant concrete plant pots integrated into the structure. The building, along with countless others, positions nature as something under the control of humans.

The aesthetics of re-use is one of the ways in which architecture has the potential to counter rather than reinforce hubris. If hubris is understanding oneself as separate to the environments on which one depends, then the aesthetics of re-use challenges this by making present some of the systemic relations in which humans are embedded.

Consider the Waste House (Figures 2.4–2.7), designed by my colleague Duncan Baker-Brown. The Waste House has been an important precedent in establishing the feasibility of re-use in architecture, demonstrating that it is possible to make permanent public buildings

FIGURE 2.3 The Flower Tower

Source: Fred Romero. CC BY 2.0. https://commons.wikimedia.org/wiki/File:Paris_-_Tower_Flower_(24954421025).jpg

FIGURE 2.4 The Waste House, exterior

FIGURE 2.5 The Waste House, interior

FIGURE 2.6 Carpet tiles find a new role as cladding

FIGURE 2.7 Discarded video cassettes helping insulate the Waste House

from what would otherwise be waste materials. The aesthetics of the Waste House are less commented on than its technical value as a prototype. But its aesthetics matter. By this, I do not mean only what it looks and feels like, although this is part of it, but, rather, how the building works as a way of looking and feeling, a means of perceiving.

Baker-Brown describes the project as "more of a provocation than a future way to construct buildings",[18] positioning the Waste House as a form of advocacy. The building makes explicit the amount of waste produced by the building industry and beyond, acting as an educational resource within the university and in hosting visits from local school children. The provocation of the Waste House, as I choose to interpret it, goes beyond advocacy, however. It may be interpreted not just as a prompt to reconsider the building industry but also to reconsider one's place in the systems of which one is part in a more general sense.

The Waste House uncovers the stories of materials, where they came from, where they would have gone, and how they came to be here – the material flows that so easily go unnoticed even when participating in them. The vapour control layers are made from festival banners I likely would have walked past in Brighton. The building is clad in carpet tiles that I might have walked on (Figure 2.6). Sitting in the building, I am insulated by discarded denim, DVDs, video cassettes, and floppy disks – all things that have been part of my life at some point (Figure 2.7). The building can be read as one moment in the journey of its many parts through their contributions to multiple wholes, being assembled, de-assembled, and (hopefully) assembled again. In using the building, I become a part of these histories and futures, these wholes. In affording sensitivity to these relations between wholes and parts, the building offers a way of cultivating sensitivity to wider patterns of ecological relationship. In this move to an aesthetic way of reading the building, cybernetic (informational, organisational, ecological) circularities come into relation with the material circularities of re-use.

Not all instances of re-use cultivate sensitivity to parts and wholes. Thought of in terms of style, the aesthetics of re-use are open to becoming distorted and commodified, leading to the architectural equivalents of selling pre-ripped jeans.[19] There is also a need to be wary of virtue signalling, greenwashing, and uncritical attitudes to recycling. By attending to the aesthetic qualities of re-use, it is possible for architectural spaces to offer ways of developing sensitivity towards ecological patterns and one's situatedness within them. In this way, the environmental benefits of re-use are not limited to the mitigation of the harm caused by the built environment in terms of material and energy. The aesthetic qualities of re-use are also a way to counter the hubris that is one of the underlying roots of ecological crisis.

Acknowledgements

This paper has evolved from a number of presentations, including a lunchtime panel session at the School of Re-Construction, lectures and seminars at the University of Brighton, guest talks at Cardiff University and Bauhaus University, Weimar, and presentations at the 2022 Relating Systems Thinking and Design (RSD11) Symposium and 2023 conference of the International Society for the Systems Sciences. It has been informed by conversations on these topics with Tom Ainsworth, Tilo Amhoff, Duncan Baker-Brown, Joanna Boehnert, Edward Buttifant, William Cork, Marie Davidová, Rich Fairley, Jon Goodbun, Davis Mak, Marinos Mavrogenis, Dulmini Perera, Simon Sadler, Arthur Siegel, Tanya Southcott, Ben Spong, Sally Sutherland, and Jeff Turko and by the events run by Brighton's Radical Methodologies Research Group. This work was supported by the Arts and Humanities Research Council [grant number AH/X002535/1] in a project funded jointly with the German Research Foundation (DFG), titled Enacting Gregory Bateson's Ecological Aesthetics in Architecture and Design.

Bibliography

Baker-Brown, Duncan. *The Re-Use Atlas: A Designer's Guide Towards a Circular Economy*. London: RIBA Publishing, 2017.

Bateson, Gregory. *Steps to an Ecology of Mind*. Chicago, IL: University of Chicago Press, 2000. Originally published 1972.

Bateson, Gregory, and Mary Catherine Bateson. *Angels Fear: Towards an Epistemology of the Sacred*. Cresskill, NJ: Hampton Press, 2005.

Fischer, Thomas, and Christiane M. Herr, eds. *Design Cybernetics: Navigating the New*. Cham: Springer, 2019.

"Flashback: Tower Flower/Edouard François." June 25, 2012. www.archdaily.com/245014/tower-flower-edouard-francois

Friedman, Alice T. *Women and the Making of the Modern House: A Social and Architectural History*. New Haven and London: Yale University Press, 2006.

Goodbun, Jon. "On the Possibility of an Ecological Dialogue." *Making Futures*, 2019. www.making-futures.com/jon-goodbun-on-the-possibility-of-an-ecological-dialogue

Goodbun, Jon, and Ben Sweeting. "The Dialogical, the Ecological and Beyond." *Footprint* 15, no. 1 (2021): 155–166. https://journals.open.tudelft.nl/footprint/article/view/5668

Goodchild, Melanie. "Relational Systems Thinking: That's How Change Is Going to Come, from Our Earth Mother." *Journal of Awareness-Based Systems Change* 1, no. 1 (2021): 75–103. https://doi.org/10.47061/jabsc.v1i1.577

Harries-Jones, Peter. "Gregory Bateson's 'Uncovery' of Ecological Aesthetics." In *A Legacy for Living Systems: Gregory Bateson as Precursor to Biosemiotics*, edited by Jesper Hoffmeyer, 153–167. Dordrecht: Spinger, 2008.
Harries-Jones, Peter. "Understanding Ecological Aesthetics: The Challenge of Bateson." *Cybernetics & Human Knowing* 12, no. 1–2 (2005): 61–74.
Pugh, Jonny, and Eddie Blake. "Deep Re-use." In *Re-Use Pedagogies*, edited by Duncan Baker-Brown and Graeme Brooker. Abingdon: Routledge, 2024.
Sassen, Saskia. "Dressed in Wall Street Suits & Algorithmic Math: Assemblages of Complex Predatory Formations." *Proceedings of Relating Systems Thinking and Design* RSD8 (2019). https://rsdsymposium.org/dressed-in-wall-street-suits-algorithmic-math-assemblages-of-complex-predatory-formations/
Soriano, Alberto, Josina Vink, and Shivani Prakash. "Confronting Legacies of Oppression in Systemic Design: A Dialogue Series." *Proceedings of Relating Systems Thinking and Design* RSD11 (2022). https://rsdsymposium.org/confronting-legacies-of-oppression-in-systemic-design

Notes

1 On the cultural case for re-use, see Chapter 6 of this volume. Jonny Pugh and Eddie Blake, "Deep Re-use," in *Re-Use Pedagogies*, ed. Duncan Baker-Brown and Graeme Brooker (Abingdon: Routledge, 2024).
2 For an introduction to cybernetics for designers, see: Thomas Fischer and Christiane M. Herr, eds., *Design Cybernetics: Navigating the New*, Design Research Foundations (Cham: Springer, 2019).
3 Gregory Bateson, *Steps to an Ecology of Mind* (Chicago, IL: University of Chicago Press, 2000), 496–501.
4 Bateson's concerns with the relation of aesthetics and ecology have been synthesised and further developed by Peter Harries-Jones. See e.g.: Peter Harries-Jones, "Understanding Ecological Aesthetics: The Challenge of Bateson," *Cybernetics & Human Knowing* 12, no. 1–2 (2005); Peter Harries-Jones, "Gregory Bateson's 'Uncovery' of Ecological Aesthetics," in *A Legacy for Living Systems: Gregory Bateson as Precursor to Biosemiotics*, ed. Jesper Hoffmeyer (Dordrecht: Spinger, 2008).
5 Gregory Bateson and Mary Catherine Bateson, *Angels Fear: Towards an Epistemology of the Sacred* (Cresskill, NJ: Hampton Press, 2005), 192.
6 Jon Goodbun, "On the Possibility of an Ecological Dialogue," *Making Futures* (2019). www.makingfutures.com/jon-goodbun-on-the-possibility-of-an-ecological-dialogue
7 The use of "hubris" in *The Roots of Ecological Crisis* is similar to Bateson's more widely quoted, and more neutral sounding, phrase "epistemological error". Both identify ways in which humans conceive themselves separate to their environment. I find hubris the clearer and more challenging term as it helps clarify how well-meaning attempts to "solve" environmental problems can lead to ecological harms.
8 Bateson, *Steps to an Ecology of Mind*, 498, 500.
9 Jon Goodbun and Ben Sweeting, "The Dialogical, the Ecological and Beyond," *Footprint* 15, no. 1 (2021). https://journals.open.tudelft.nl/footprint/article/view/5668
10 Bateson, *Steps to an Ecology of Mind*, 497.
11 Systems thinking is itself not immune from this. See e.g. Melanie Goodchild, "Relational Systems Thinking: That's How Change is Going to Come, from Our Earth Mother," *Journal of Awareness-Based Systems Change* 1, no. 1 (2021). https://doi.org/10.47061/jabsc.v1i1.577; Alberto Soriano, Josina Vink, and Shivani Prakash, "Confronting Legacies of Oppression in Systemic Design: A Dialogue Series," *Proceedings of Relating Systems Thinking and Design RSD11* (2022). https://rsdsymposium.org/confronting-legacies-of-oppression-in-systemic-design
12 Bateson, *Steps to an Ecology of Mind*, 498.
13 Saskia Sassen, "Dressed in Wall Street Suits & Algorithmic Math: Assemblages of Complex Predatory Formations," *Proceedings of Relating Systems Thinking and Design RSD8* (2019). https://rsdsymposium.org/dressed-in-wall-street-suits-algorithmic-math-assemblages-of-complex-predatory-formations/
14 Alice T. Friedman, *Women and the Making of the Modern House: A Social and Architectural History* (New Haven and London: Yale University Press, 2006), 138.
15 Mies van der Rohe in conversation with Christian Norberg-Schulz, quoted in Friedman, *Women and the Making of the Modern House*, 139.

16 Friedman, *Women and the Making of the Modern House*, 138.
17 "Flashback: Tower Flower/Edouard François," June 25, 2012. www.archdaily.com/245014/tower-flower-edouard-francois
18 Duncan Baker-Brown, *The Re-Use Atlas: A Designer's Guide Towards a Circular Economy* (London: RIBA Publishing, 2017), 145.
19 Pugh and Blake, "Deep Re-use."

PART III

Themes pursued during the International School of Re-Construction

3

RAW 1 'SOCIAL FABRIC'

Nick Gant and Ryan Woodard

FIGURE 3.1 Waste brick illustration

Source: Nick Gant and Ben Bosence

Social fabric studio

Our small team convened across countries and cultures from locations including China, Belgium and sites across the UK. We developed a perspective of re-construction that envisions opportunity through a somewhat 'Marx-ish' notion of valorisation of waste through the application of labour. However, this process is restorative and reinvigorating in its ambition to rewrite material narratives that manifest through the build environment. These stories form through

DOI: 10.4324/9781032665559-8

This chapter has been made available under a CC-BY-NC-ND 4.0 license.

past-precedents, 'parallel-presents' and affirmative-futures that re-construct the way communities engage with materials as a means to, literally, make change locally.

Introduction

Through this studio we *rethink* the relationship between people and place as mediated by waste materials and processes of physical and philosophical *meaning making*. As a studio of individuals from different localities, we undertake a collective design process and articulate and illustrate a programme of exploration and reflection through material maps, images, illustrations and made objects related to the materiality of place. We consider the social value of materials (that would otherwise be waste) in our neighbourhoods and contemplate their role as arbiters between, and signifiers of, different cultural value systems. Using *material as evidence,* we document a form of anthropological archaeology, a process that *reveals*, *re-forms* and *re-presents* the social, the sub-cultural and the material behaviours of people in their place. It (sometimes literally) casts new material narratives through the physical and metaphysical social fabric that surrounds us by engaging with waste as *culturally loaded matter*.

This studio and chapter explore tried-and-tested, incremental and accumulative design processes that document socio-material systems and participatory processes. We map the environments in which our social-design 'actors' and participants live and the waste(d) opportunities that provide latent potential for social stories and opportunities that are afforded through more circular modes and methods of meaningful making. We chart a course through neighbourhoods using the prism of waste material and waste(d) opportunities identifying potential 'spin-out' benefits for social and cultural value and renewed potential for more 'virtuous-circular-economies' and enhanced 'material literacy' [1]. The result is (more) meaningful-matter, material *products* that embody something of their *place* in relation to its *people*.

Context

Lynch [2], a leading environmental design theorist, provided two visions of what future society will look like in his study 'Wasting Away'. '*Waste Cacotopia'* sets out a scenario where the linear 'make, take, dispose' economy prevails. A world where we have exhausted Earth's resources, and we have become dependent on importing materials and resources from the Moon and other planets. Nations fight over land for waste disposal and release toxic gases contained in plastic bags which float out and dissipate into deep space – a form of space junk that, several decades later, is becoming a real-life challenge [3]. In contrast, *'Wasteless Cacotopia'* embodies a world where *extreme* circular thinking has been deployed. A waste-free society with clean air, no more sewage, an unencumbered Earth, producing optimal environmental outcomes. Uniformity and predictability are employed in the design and function of all consumer items resulting in a uniform urban environment. Plants and animals would be bred to reduce *useless* parts: '*stringless beans and skinless beets and boneless chickens*'.

Whilst we are striving for a *'Wasteless Cacotopia'*, be it not one so radical as Lynch's vision. Circular thinking is becoming increasingly embedded in government policy and industry business models, and the prevalence of resource management is well documented throughout human civilisation. As far back as 1 CE, Pliny the Elder referred in his writings to the re-use of scrap metal [4]. In ancient Rome, it was common for old buildings to be dismantled and stones reused in new constructions. This was not just for logistical reasons but also aesthetics. A famous example is the Arch of Constantine in Rome, which incorporated fragments from earlier monuments to Trajan and Marcus Aurelius [5]. There were extensive networks in place for collecting faeces to be sold to farmers as 'night soil' fertilizer and urine for dyeing fabrics.

In the UK, we don't need to look back too far to revisit a time when re-use and recycling was a key part of daily life as an important contributor to the economy and the material fabric of our built environment. In London, by the beginning of the 1800s, there was an effective system for collecting and processing waste, which was driven by the economic value of the *resources* being discarded. There were multiple stakeholders – some whom specialised in collecting certain materials such as dog waste, which was used in the tanning industry to preserve leather, and rag-and-bone men who remained commonplace in U.K. society until the latter 20th century and popularised in the sitcom *Steptoe and Son*. The rags were sold for paper-making, and bones had fat and marrow removed, which were sold for soap or glue-making with the remaining crushed bones sold as a fertilizer [6]. Conversely, there were enterprises, which collected mixed materials to be deposited at dust yards, a primitive version of today's materials recovery facilities that we see in place today. Here, predominately women and children, working in unsanitary conditions, would segregate materials to be sold into the respective markets. By far, the largest component of the waste stream, 53% by weight, was fine dust and ash [6]. This was generated from open fires being used for cooking and to heat homes – and the term *dustbin* originates from its abundance in the waste. The market for dust and ash was thriving due to the demand from brick manufacturers. The population of London increased by 145% within a 50-year period, with that the demand for new housing exploded. Dust was in plentiful supply and made for a cost-effective alternative for local materials that were becoming diminished. The material realisation of a social boom was embodied with that same community's waste – moreover, social and economic benefit was derived from the valorisation of waste. However, like all commodities, value is based on supply and demand, and by the late 1800s, the demand for dust drastically reduced due to the widespread excavation of clay, which was seen as a more durable and viable material for brick making. For example, in 1878, the parish of Kensington generated £2,318 from the sale of dust (the equivalent of £218,370 in today's value), but by 1886, this had reduced to nothing as the market collapsed [7] and thus the socio-economic and material landscape of the city changed.

With the increasing emphasis on environmental protection and advances in new technologies, dust is being revisited as a resource to build with. For example, in India, bricks have been developed utilising 80% dust and 20% plastic waste, thereby utilising waste material and reducing CO_2 emissions from using conventional materials [8]. In neighbouring Nepal, Kathmandu Metropolitan Council has unveiled plans to convert the 17 tonnes of dust collected each day from road sweepers into bricks [9]. Companies are developing technologies to 3D print buildings utilising waste streams and biocompatible materials. A leading proponent is the Italian company WASP that has developed 3D printed houses and was a collaborator in the realising of the 'House of Dust' installation based on the 1967 computer-generated digital poem of visual artist and founding member of the Fluxus movement Alison Knowles. The poem was the basis for an interactive sculpture on the California Institute of the Arts campus in the early 1970s, which was recreated in 2021 through 3D printing over 60 cubic meters of natural material [10].

It's interesting to reflect on Lynch's '*Waste Cacotopia*' vision and the dependency on resources from other planets. Both the European Space Agency and NASA have a vision of setting up a permanent base on the Moon – for example, NASA's Artemis programme aims to establish a base camp that includes a modern lunar cabin, rover and mobile home. Rather than transporting construction materials from Earth, scientists have started to develop prototype building materials harnessing lunar dust, Martian and Lunar regolith [11]. The surface of the Moon is covered in grey, fine, rough dust – estimated to have a 40% oxygen content. The malleable properties of the dust means it can be crushed, burned and compressed making it an ideal material for brickwork. It can be utilised as feedstock for 3D printing buildings [11], and NASA's 3D Printed Habitat Challenge was designed to advance the construction technology needed to create sustainable housing solutions for Earth and beyond [12].

It's a cliché, but *waste* streams should be seen as *resource* streams. Even the low-base materials such ash and dust – perceived to have no *value* economically and socially – have been key resources in the development of infrastructure over centuries and will seemingly be pivotal in the social and environmental fabrication of our places in this world or others in the future.

Reconstruction method 1: making maps

In each remote location, we undertook 'walk-shops', simultaneously seeking out waste opportunities that we mapped using the Community21 open mapping system:

- What are the abundant materials present in our redundant buildings or by-products and that populate and personify our particular neighbourhoods?
- Who and what are the people and places that could be co-opted to supply reconstruction materials and services?

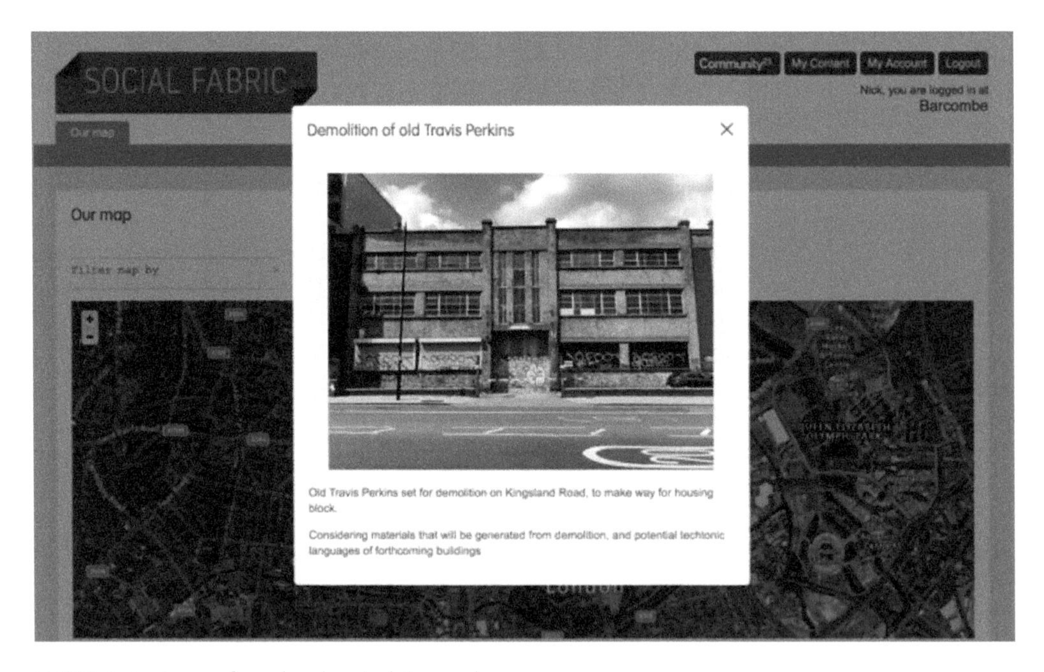

FIGURE 3.2 Image from local material mapping

Reconstruction method 2: making meaning

Our *Social Fabric Studio* considers a modern vernacular that is defined by locally authenticated materials that have 'substance' in terms of their physicality and their ideology, their materiality and their meaning. The built form in this context reinterprets the observations of Helena Coch, referring to a more 'representative' architecture; 'popular' in its authentic reflection and *account of climate change*, being less pretentious in use of both locally resourced, artificial, and natural materials [13]. This type of *accounting for climate change* as a relationship between material and site [14] manifests through the actual material use, in an attempt to negate further emissions, to retain finite elements but also to define an identifiable material language as '*mediating-matter'* [15]. Material compositions that function as a means to form the build spaces that we need but that communicate both *issue and opportunity* as a kind of product-protest, poetic and politic. It is publically presented by, and through, abundant, uncultivated materials that are remade. Our objects may present a kind of material characterisation of consumption and consequential re-use of locally 'abundant' (potentially waste) materials – something that is common to vernacular architecture through time [16].

These materials are often considered to be regionally specific and culturally distinct with potential for a rich diversity of sub-cultural variation *baked in* through local accessibility and availability and as a reflection on society – objects and buildings as 'cultural facts' [17]. Being resourceful is a form of activism (in our current context). Re-use is helping to reduce the present peril that faces all communities – therefore, vernacular is social in its incarnation and acknowledgement of local need, resilience and adaptation, evolving more direct and actionable responses to environmental forces. Socially and culturally, *re-mining re-sources* (and their meaning) becomes a moderate material movement and an act of aesthetic activism – this we seek to aestheticize and materialise in visceral, readable forms. Talking through tactility [18] comprehending 'architecture as a vehicle of communication through its material state, becoming an explicitly curated, visual-material record of our time and therefore a reflection or image of life and culture in the street' [19].

FIGURE 3.3 Recycled block illustration

Source: Nick Gant and Ben Bosence

Two sides of the circle

So in our Social Fabric Studio and scenarios, we assert the notion that there are 'two sides to the circle' in circular economies (see Figure 3.4) – one is the technical management of the material; objectively and scientifically assessed and governed by (waste) materials as quantifiable 'stuff' that is to be calculated and mechanically transformed as a physical commodity. The other side recognises that waste may result from what we 'think and feel' as well as what we do and, therefore, we need to engage with reconstruction materials as meaningful stuff that is embodied with culture and history, infused with interactions with us and forming part of our constantly changing behaviour and our values. This goes beyond the numerical through material relationships between our lives, our consumption and the world around us. Therefore, as re-constructors, we must recognise both the objective and subjective, the physics and the poetics, if we are to engage in persuasive and 'sustainable' acts of change.

By embracing the unpredictability of behaviour, new opportunities arise for diverse interpretation, variation and expression beyond reforming landscapes monopolised by the monotony of mundane, material monoliths. There is enormous scope to create diverse outputs through the creative application of waste and in changing the perception of waste [20].

Reconstruction method 3: remaking (meaning) machines

> *Meaning is inserted at the point of making the literal, material and existential fusion of skill, knowledge and ideals into an everyday type of artefact* [21].

Technologies emerge from a response to conditions and local forces [22] and our Social Fabrication materials and making methods reflect this within the machines that help us 'make it' (our more socially and materially resilient future). The actuation of our visions for alternate means-to-more-resourceful-ends

In the reconstruction scenario, we considered how this kind of 'Space-crafting' [23] addresses fundamental need for buildings fabricated with local materials whilst defining new opportunities for a situated and distributed economy. We engineer open-source re-mining machines designed to

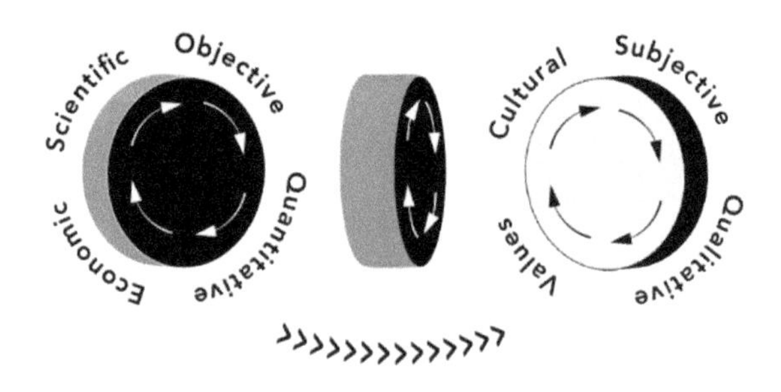

FIGURE 3.4 The two sides of the circular economy

Source: Nick Gant

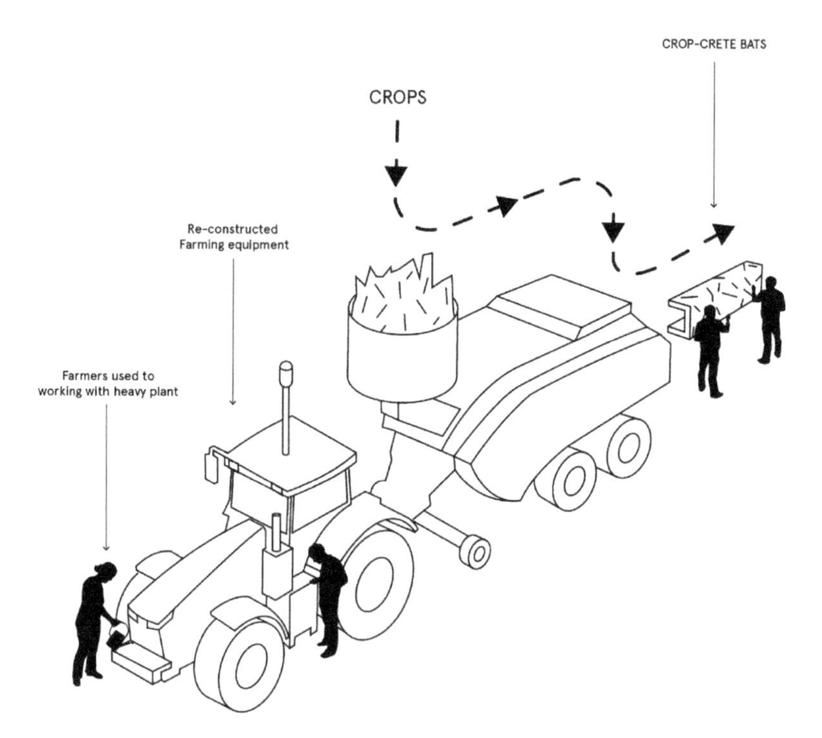

FIGURE 3.5 Re-mining machines/re-purposed farming equipment generating crop-crete bats for social housing.

Source: Nick Gant

promote forms of multi-local-manufacture, where materials remain localised and enriched through their re-use and ideals and ideologies are shared beyond the local to promote diverse (sub)cultural value and material discourses and dialects. Through *social fabric-action*, localities become centres for reinvestment in *people, place* and the *products* of those processes. Local infrastructure is reconstructed to favour localised remanufacture between *knots-in-a-network* of distributed intelligence and know-how. Reconstructed outcomes encompass modest, personalised, bespoke (re)incarnations by *have-a-go heroes* as well as the systematic re-routing and re-mapping of entire cities and regions, all of which enact self-centred-systems and infrastructures of hyperlocal re-production. Re-production is re-imagined shifting from the inauthentic representation of past archetypes and disingenuous, faux facades into realisations that faithfully reform materiality into new typologies of change and transition. RE-IY (recycle yourself) is embodied with the value and meaning of personal, local investment; self-actuated adaptation; resourcefulness and transformation.

Reconstruction method 4: Parallel Presents

We undertook a studio speculative design workshop adopted and adapted as a version of Amy Holroyd Twigger's 'Fashion Fictions' to form 'Parallel Presents' [24]. We generated individual visions for material uses in imagined parallel worlds that are inspired by and/or reflect our own alternatives and opportunities for more sustainable ways of being in the world. These were briefed around our ideas and ideals for meaning-making through the re-application of waste in socially and environmentally resourceful ways.

Examples of our ‘Parallel Presents’ imagined alternates:

People’s homes are built to meet all of their needs in housing developments across the UK. The origins of this culture can be traced back to the revocation of the Right to Buy scheme in 1980. Developers work with local tradespeople in the area, up-skilling them in how to resource map and turn these resources into a reusable product. People provide their own waste to be used in the building and have agency throughout the process. Developers learn to listen to future residents and work with them to build something to last.

(Olivia)

All public buildings are mixed used, restored and never fully ‘new-build’ in the UK. The origins of this culture can be traced back to resource scarcity and de-globalisation due to COVID-19 pandemic and 15-minute city campaigns. As a result of this practice, heritage, history and locality are more valued. Sustainability and circularity in design are the norm rather than a plan. Buildings tell a story and become a timeline of the place they occupy.

(Bruna)

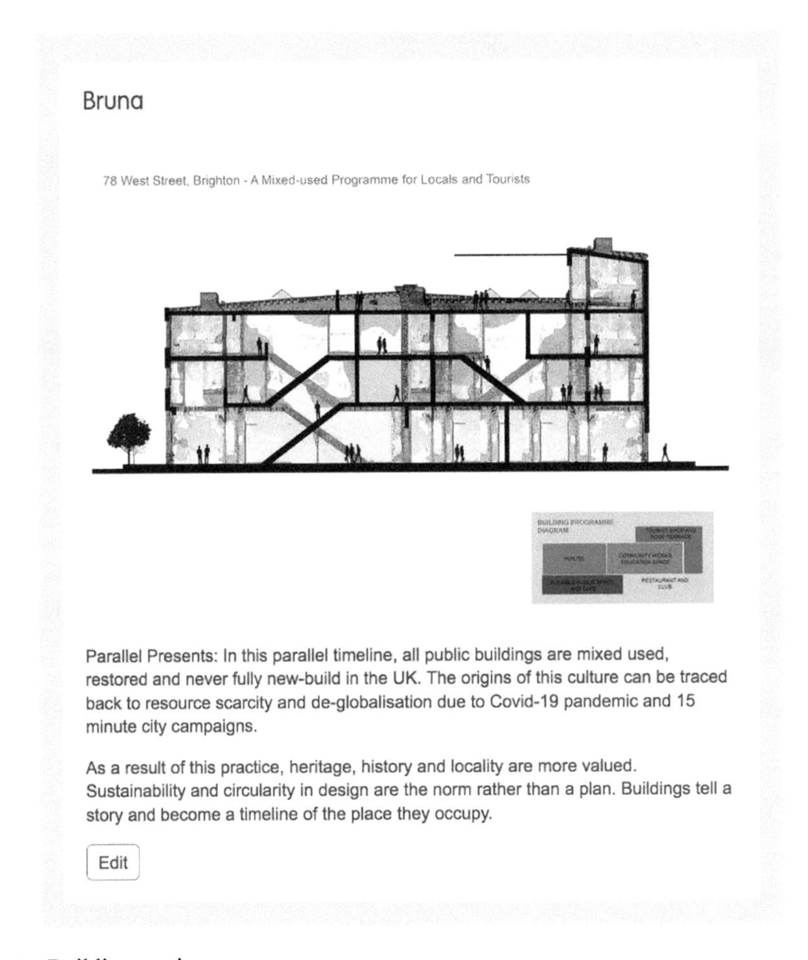

FIGURE 3.6 Building section

Source: Bruna Borges

Bruna

Parallel Presents: In this parallel timeline, all public buildings are mixed used, restored and never fully new-build in the UK. The origins of this culture can be traced back to resource scarcity and de-globalisation due to Covid-19 pandemic and 15 minute city campaigns.

As a result of this practice, heritage, history and locality are more valued. Sustainability and circularity in design are the norm rather than a plan. Buildings tell a story and become a timeline of the place they occupy.

Edit

FIGURE 3.7 Social Fabric actors

Source: Bruna Borges

Container fab-labs are built in Belgian recycle parks as a make-place to reuse construction and demolition waste and turn it into street furniture, tiles . . . The origins of this culture can be traced back to the decree of waste in 1981. Companies such as MIWA organise recycling parks in such a way that citizens collect and drop off their resources they want to get rid of and recycled. But in 1 of the fab-lab containers you will find local up-skilled instructors making something innovative with it regarding the need in the city. Having less waste, less container-transport, less pollution.

(Julie)

Buildings last a thousand years. The origins of this culture can be traced back to a moratorium on new construction, forcing a rethink of the Approved Documents in 2033 (IPCC).

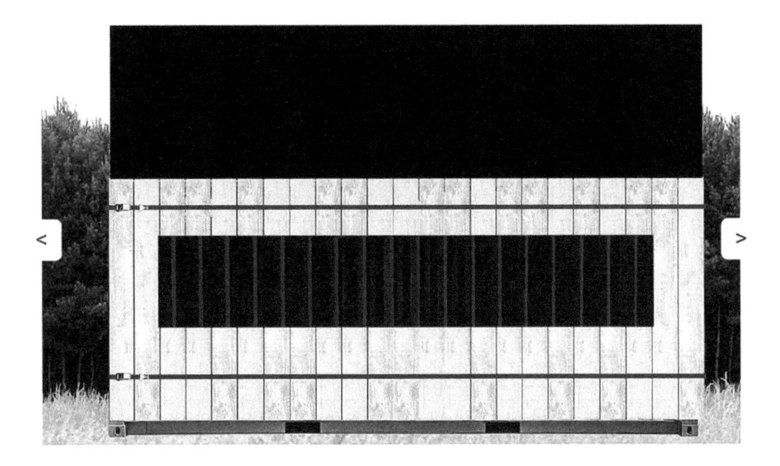

FIGURE 3.8 Reconstruction building elevation

Source: Leo Sixsmith

> *Buildings must be repaired and maintained. In order to be economical, new components are robust but easily removable, and a new rental market of building components flourishes. Building tectonics become much more legible as a result. The freedom this offers enables the particular elements of buildings to be easily adapted.*
>
> *(Leo)*

These stories provide an insight and basis for action and reaction, reconstruction of vernacular language and material hierarchies – means to endless ends, where the circulation of stuff establishes new opportunities as social substance.

Social Fabric Studio summary

Through our individual, place-based-perspectives on waste, our 'Social Fabric' studio explored new potentials to reframe re-use and re-construction as a mediator of meaning that is socially and culturally significant in its representation of us, who were are, what we are, what we do and how we do it. In each location, we remotely 're-mined' the material and human re-source to uncover new (or latent) opportunities for re-constructing our physical-places and social-spaces revealing emergent forms of locally authentic making and manufacture. As a result, we rewrite our 'multi-local' material narratives and reform our interconnections as a means to redefine and recast the role of designers and makers as actors and curators of new social, cultural and environmental eco-systems and services – all of which manifest through the application of meaningful matter(s). Our buildings can be read as scripted visual verse, positive pros, annotating our material sub-cultures through idiosyncratic, localised languages of reapplication and reiteration. Any reinstatement of a re-built form punctuates the long lives of materials that pass through time in the service of communities' needs – surfaces and sections form the layered strata that tier up narratives of use and re-use of materials in space and time. The material recipes for composition reflect each localised community and context in

FIGURE 3.9 Local reconstruction

Source: Yihang Wang

terms of cycles of (more) considerate consumption and indigenous means-of-making. With reinvestment in hyperlocal production comes greater material literacy [25] throughout the community, understanding the production, provenance and proficiency that 'makes-up' the build environment. Nuanced distinctiveness of local material curation, configuration and communication becomes deeply embedded in the cultural constitution and sub-cultural variability that emerges out of former, synthetic homogeny and monopolies of buy-and-bury-or-burn behaviours of the past. Local material archetypes are re-authored through literate eco-logical idioms, erudite material mash-ups and positive propaganda. Through the programme, each of our studio participants was able to identify and author opportunities for material meaning making and define ways in which the physical, eco-nomic and sub-cultural landscape can be redefined. In each case, opening opportunities to re-establish material cultures that can re-invigorate methods for *social fabrication* enacted and afforded by diverse and locally distinctive methods for re-construction.

Thanks to our Social Fabric Studio participants/students Olivia Harrison, Yihang Wang, Jason Wan, Leo Sixsmith, Bruna Borges and Julie Van Raemdonck.

Reference List

1. Gant, Nick. (2017) Mediating matters. In *Routledge Handbook of Sustainable Product Design*, pp. 222–235. Oxford: Routledge.
2. Lynch, K. (1990) *Wasting Away*. Sierra Club Books, San Francisco.
3. Witze, A. (2018) The quest to conquer Earth's space junk problem. *Nature*, vol. 561, no. 7721, pp. 24–26.
4. Mazzatenta, L. (1995) Classical castoffs reclaimed from the sea. *National Geographic*, vol. 187, no. 4, pp. 88–101.
5. Medina, M. (2007) *The World's Scavengers: Salvaging for Sustainable Consumption and Production*. AltaMira Press, Lanham.

6. Velis, C.A., Wilson, D.C. & Cheeseman, C.R. (2009) 19th century London dust-yards: A case study in closed-loop resource efficiency. *Waste Management*, vol. 29, no. 4, pp. 1282–1290.
7. Gordon, W.J. (1890) *How London Lives: The Feeding, Cleansing, Lighting and Police with Chapters on the Post Office and Other Institutions*. The Religious Tract Society, London.
8. Kothari, M. (2022) Circular economy through resource recycling: Creating wealth from industry & plastic waste. *Machine Maker*, 5th April 2022. https://themachinemaker.com/market/circular-economy-resource-recycling-0604. Accessed 28th February 2023.
9. Ojha, A. (2019) Metropolis plans to make bricks from dust collected from the streets. *The Kathmandu Post*, 2nd May 2019. https://kathmandupost.com/valley/2019/05/02/metropolis-plans-to-make-bricks-from-dust-collected-from-the-streets. Accessed 5th February 2023.
10. WASP (2021) Crowdfunding for the house of dust. *WASP*, 1st March 2021. www.3dwasp.com/en/crowdfunding-for-the-house-of-dust/. Accessed 5th February 2023.
11. Warren, P., Raju, N., Ebrahimi, H., Krsmanovic, M., Raghavan, S., Kapat, J. & Ghosh, R. (2022) Effect of sintering temperature on microstructure and mechanical properties of molded Martian and Lunar regolith. *Ceramics International*, vol. 48, no. 23, pp. 35825–35833.
12. NASA (2021) *STMD: Centennial Challenges*, 12th January 2021. www.nasa.gov/directorates/spacetech/centennial_challenges/3DPHab/index.html. Accessed 5th February 2023.
13. Coch, H. (1998) Bioclimatism in vernacular architecture. *Renewable and Sustainable Energy Reviews*, vol. 2, no. 1–2, pp. 67–87.
14. Weber, W. & Yannas, S., eds. (2014) *Lessons from Vernacular Architecture*. Vol. 2. London and Oxford: Routledge.
15. Gant, N. (2017) Mediating matters. In *Routledge Handbook of Sustainable Product Design*, pp. 222–235. Oxford: Routledge.
16. Asquith, L. & Vellinga, M., eds. (2006) *Vernacular Architecture in the 21st Century: Theory, Education and Practice*. Oxford: Taylor & Francis.
17. Glassie, H. (2000) *Vernacular Architecture*. Vol. 2. IA: Indiana University Press.
18. Edelkoort, L. & Fimmano, P. (2021) *A Labour of Love*. Lecturis, Eindhoven.
19. Attfield, J. (2000) *Wild Things, the Material Culture of Everyday Life*. London: Bloomsbury.
20. Bramston, D. & Maycroft, N. (2014) Designing with waste. In *Materials Experience*, pp. 123–133. Oxford: Butterworth-Heinemann.
21. Attfield, J. (2000) *Wild Things, the Material Culture of Everyday Life*. London: Bloomsbury.
22. Coch, H. (1998) Bioclimatism in vernacular architecture. *Renewable and Sustainable Energy Reviews*, vol. 2, no. 1–2, pp. 67–87.
23. Sleumer, N. (2020) Re-miming – a technological renaissance out of human and mineral exstraction? In J. Boelean & M. Kaethier (eds) *Social Matter, Social Design*. Valiz, Amsterdam.
24. Holroyd-Twigger, A., 23rd May 2022. https://fashionfictions.org/ accessed.
25. Gant, N. (2017) Mediating matters. In *Routledge Handbook of Sustainable Product Design*, pp. 222–235. Oxford: Routledge.

4

RAW 2 'REIMAGINING INFRASTRUCTURE'

Scott McAulay and Sam Turner

To design for the uncertain, warming future that is rushing fossil-fuelled towards us, buildings can no longer be considered as individualistic, isolated objects. By 2050, much of the world's physical infrastructure – bridges, dams, flood defences, housing, motorways, energy grids, reservoirs, sewers, storm drains, and far more besides (notably those built during post-WW2 re-construction efforts) – shall be close to a century old; effectively, they are dependent upon reserves of resources and have been designed for worlds that no longer exist. Modernity's default of reinforced concrete skeletons – hosting and holding up its extractive, energy-generating, or transportation systems – are nearing the end of their lives. Before, and after, this ageing infrastructure fails, how might we intervene, what other purpose could it fulfill, and what comes next? At the glacial pace society is moving to secure an equitable and liveable future, time is running out to do those questions justice and to mobilise effectively around them.

Our civilisation's infrastructure and the injustice of its distribution shape life as we know it.[1] It is the technosphere's mechanistic mycelium: single-function arteries, cables, tendrils, and roots, running above, along, and below the Earth's surface, (re)connecting us whilst separating us from nature,[2] facilitating neoliberal capitalism's separation of individuals from one another,[3] and polluting our Earth to keep the lights on. It also enables, locks us into, and sustains the Anthropocene as we experience it, offering access to purchasable, drinkable water and warmth in our homes and, if you can pay for it, the transportation of ideas, energy, materials, and even worldviews, around a world colonially terraformed.

How might the infrastructure of the Anthropocene be used more regeneratively as climate breakdown accelerates its redundancy? What forms might infrastructure capable of sustaining a circular, decarbonised civilisation – one not reliant upon gigatonnes of concrete[4] to destroy, dominate, or suppress nature[5] – take? And what forms of social infrastructure might need cultivation to support a Just Transition that gets us there? These are questions that we must make time to answer imminently: so, let us begin.

DOI: 10.4324/9781032665559-9
This chapter has been made available under a CC-BY-NC-ND 4.0 license.

Radical imaginaries beyond buildings

To imagine and work towards safer and more just futures, architecture students must be empowered, taught, and inspired to think beyond the building-by-building scale and begin to learn about interventions on infrastructural and systemic levels. What any given civilisation builds locks it into certain ways of living for decades at a time – especially the infrastructure that enables its habitation – and we cannot keep going as we are without locking in ever more catastrophic climate consequences.[6] Based on the questions we ask of it now, infrastructure will either be an enabler of climate action over the next few critical years or it shall continue to be a colossal hindrance.

Despite humanity's technological advances, global energy demand is still rising,[7] increasing strains on ageing electricity grids and driving demand for more energy generation infrastructure to keep pace. This raises questions of equity and justice, as very few countries can afford to increase generation capacities indefinitely, and such – often polluting – infrastructure is often placed near marginalised, racialised, and working-class communities. At the same time, countries that can shoulder such costs – such as the United States, the EU-27, and the UK – were the largest contributors to the crisis we now face,[8] and owe significant loss and damages, and further reparations, for historical responsibility for emissions.[9]

Whilst we hear about carbon evermore often in architectural spaces, we must not lose sight of the fact that radically reducing the wastefulness of the construction sector and our physical civilisational infrastructure is essential to realise a circular economy and to decarbonise any country. After all, reaching net zero carbon is not just about reducing emissions; it is about saving millions of lives, and measures like energy efficiency and renewable energy can only get us halfway there. The other half is down to materials and societal circularity.[10]

Making matters of infrastructure even more interesting, in 2021, the International Energy Agency (IEA) confirmed that there can be no new oil, gas, or coal development if reaching net zero carbon by 2050 is to remain achievable.[11] Adhering to this would halt the growth of, and necessitate the phase-out of, the fossil fuel industry and spark radical changes to energy systems around the world. Meaning that depending on how an energy transition is managed, the entire industry's infrastructure could either become available resources or (literally) stranded assets. The IEA's announcement has emboldened campaigns to transition away from fossil fuels without abandoning the industry's workers and supercharged civil resistance as governments, like the UK's and Norway's, intend to maximise extraction despite that scientific consensus and intensifying extreme weather events. Overlooking infrastructure – of any single kind, be it energy, water, or otherwise – is not an option as we fight for a Just Transition towards a more compassionate, circular, and regenerative society.

Welcome to the Anthropocene

Projections (from bodies like the Intergovernmental Panel on Climate Change[12], and the World Meteorological Organisation[13]) of the Earth's changing climate, the damage climate breakdown shall do, and the suffering that will result offer stark visions of possible futures. Temperature records being broken and extreme weather events that devastate and end lives are now experienced year-round, and seen on our screens daily.[14] As I wrote this chapter, the United Nations observed the hottest week on record.[15] Our situation is dire, and it cannot help matters that there are significantly more works of cinema, fiction, and television offering apocalyptic or dystopian stories of climate change than those where we adapt and flourish despite it.[16,17]

Scientific consensus is that human activity is now the greatest force acting upon the Earth's planetary systems: taking us out of the Holocene – the geological period that provided the stability and seasonal regularity in which humanity evolved and established civilisation as we know it – and into the Anthropocene.[18] Of all the materials extracted from the Earth each year, 27% now goes towards the net addition of buildings and infrastructure,[19] fuelling the construction of a floor area equivalent to the city of Paris every single week.[20] In 2020, human-made materials surpassed 1,100 gigatonnes and began to outweigh all living and naturally occurring matter on Earth,[21] putting the magnitude of the ecological crisis we are living through into a sharp, jarring focus, that is staggeringly hard to comprehend or process.

As the climate crisis escalates, climate anxiety, dread, and grief become more prevalent. A *Lancet* study published in 2021, involving 10,000 young people between 16 and 25, revealed that 75% of them think that the future is frightening and 83% think that people have failed in taking care of the planet.[22] Intensifying climate impacts result in accumulative trauma where they are experienced as well as in those who learn of them, and wrenching our thoughts away from dystopian imagining can be difficult.

But what if – as part of fighting for a better world – we were to collectively reimagine and broaden our definition of infrastructure so that we might intervene and transform what a changing climate shall render redundant, in such ways that life might thrive in 2050 despite intensifying climate breakdown? What might that look like, and how might that work? This was the invitation that Anthropocene Projects extended to our team for the School of Re-Construction.

Educating otherwise for a warming world

Typically, contemporary architectural education encourages post-rationalisation, saviourism, and, in worrying cases, the obedient fulfilment of project briefs that have been recycled for years on end – rarely acknowledging the extent or existence – of climate breakdown. In many design studios, students are encouraged to decide what "problem" they wish to "solve" whilst meeting a rigid project brief. Often work is situated in an area students visit once or twice during class hours, without familiarisation, consistency, nor consent to be working with those they wish to "help". Statistics are then used to retrospectively justify their approach. Instead of recognising and building an understanding of the fact that much ecological harm and societal struggles originate in degenerative local, national, and global systems, students are encouraged to prescribe a (new) building to solve this individual "problem" whilst taking guidance from tutors who hold power over their grades. Therefore, we began our workshop by leaving such traditions in the past and grounding the studio in the reality of a climate crisis and our students' lived experiences.

Before meeting our students for the first time, our studio brief laid out a pair of complementary research tasks that would lay the groundwork for each student's eventual outputs for the summer school and form the basis of our first session. Very consciously, Sam and I (Scott) invited the students to think into the future and to delve into predictions for 2050 – using and citing reputable scientific data rather than rooting their thinking in the present day. We were careful not to prescribe a format, medium, prompt in any direction of our choosing, or to define a specific scale for these place-based explorations; this opened space for imagining beyond traditional notions of infrastructure and scales of architectural intervention. We wanted the 10-day summer school to be an opportunity for our students to focus on something that captured their imaginations – perhaps something that they could develop further to cultivate their climate literacy – and that gave them a greater sense of what is already possible in terms of responding to climate change. As such, we made it very clear that they would get as much out of the fortnight as they put in, and that, unlike traditional schools of architecture dynamics, we would be there to facilitate and support their explorations, investigations, and learning rather than to lead, specify, or shape outcomes.

As things stand, Earth is set to surpass the 1.5 °C ceiling for global heating – the threshold set by the Paris Climate Accords[23], so we set our studio in the year 2050, in a world experiencing 2 °C of global heating. Such an average temperature increase would render hundreds of millions of buildings unable to safely mediate their internal environments during heatwaves without extensive adaptation. It would also erase vulnerable island communities, many losing areas of cultural significance in the process, and aggressively redraw coastlines the world over, forcing millions to relocate and rendering much infrastructure useless. In such a world, stranded assets will no longer be confined to economic terminology: it will become a lived reality of communities around the world. Rising sea levels and intensifying wildfires will render vast tracts of land, and the homes upon them, uninhabitable, and much infrastructure – including all the resources locked up in them – will be left behind new shorelines, with their salvage becoming increasingly difficult, costly, and – quite ironically – more resource-intensive over time.

Bearing this in mind, the students' first task was to investigate, and to question, how global temperatures rising by 2 °C would affect the region in which they live, using scientific studies and reports. With students based in Belgium, Hungary, China, Portugal, Greece, and England, not only did student's outcomes vary – allowing for rich discussions, but the exercise offered them insights into how climate change is affecting different parts of the world in a participatory way rather than being lectured. We set no parameters nor strict methodology, leaving this to individual interpretation, and it set the scene for the landscape into which each student would place their work.

With the School of Re-Construction taking place in August 2021, extreme weather was fresh in the memories of our students. Belgium had experienced unprecedented flooding the previous

FIGURE 4.1 Satirical, propaganda-esque visions of Athens' climate future without adaptation – drawing from tones of climate denial in political discourse

Source: Marvina Sinjari

month that its infrastructure was not designed to cope with; the south of England's ageing flood defences had been overcome by torrential rain; and Athens (where one of our students was living) experienced heatwaves and ferocious wildfires (an occurrence that is repeating and intensifying in 2023 as I write this).

The implications of a 2 °C world on water management and how we live with water became a key theme and focus for some students very quickly. They noted sea level rise as being an imminent threat to places they know well, intensifying precipitation cycles leading to flooding and disruption to food production, evolving approaches to living with rather than fighting against water ways, and even the staggering amount of water being used to cool Belgian nuclear power plants in 2021. A second theme from the students' investigations was extreme heat: overheating was a worsening issue in densifying cities and urban centres that have become "concrete jungles" with little greenery, particularly cities like Athens (Figure 4.1) where modern buildings have not been designed for passively cooling and, instead, rely on mechanical measures, and the resultant wildfires that had been raging out of control across Europe, at just 1.1 °C above pre-industrial levels. Beyond these themes, discussions were had about impacts on mobility amidst displacement and losses of biodiversity and nonhuman life due to climate change.

This first task essentially took the place of "traditional" site analysis but fused it with learning about how climate change is going to affect people around the world, the places we know well, and the buildings and infrastructure around us. What stood out is that whilst some students took a step back and investigated things at the national level, others zoomed in and identified sites in deindustrialising or abandoned areas, began to compile precedents of successful mitigation measures, and developed satirical posters to communicate the absurdity of political responses to, and rhetoric about, what was blatantly climate change in action (Figure 4.2).

The second research task invited students to identify unloved buildings, raw materials, or infrastructure around them that would become useless in that 2 °C world. Ways of designing, building, and living today that have become a comfortable, expected norm – for a privileged percentage of humankind – are grossly unsustainable and depend on ever-larger sacrifice zones to exist. Sacrifice zones being geographical areas in which typically poorer, marginalised, and racialised peoples are exposed to dangerous chemicals, environmental stressors, and pollution from harmful industrial activities.[24] Construction is a huge contributor to this.

Our students' approaches to the second task differed wildly, which led to some incredibly rich conversations and exchanges of knowledge. Our two students in Belgium both opted to focus on reimagining housing along the country's threatened coast as material banks but came back with radically differing approaches to what their material might be and how those houses might be used: one imagining them as repair kits for other buildings and the other as resources to build elsewhere – further inland, in dynamic, nomadic, and less static circumstances. Coincidentally, across China, Hungary, and England, three of our students were drawn to abandoned industrial complexes – within cities and far beyond them, envisioning them as the start of something new, such as a new purpose in building resilience to climate breakdown and the more extreme weather it will bring. Our last two students became fascinated with alternative uses for and giving new life to specific items: housing lying incomplete since a financial crash; air conditioning units, countertops, and sinks made from precious stone in Greece; and remnants from the Pedogão Grande forest fire in Portugal.

Complementing these research tasks was background reading and listening resources focussing on climate solutions, stories of the possible, and visions of possible futures. Sam suggested students

FIGURE 4.2 A second satirical, propaganda-esque vision of classical architectures struggling to cope in a climate future without adaptation

Source: Marvina Sinjari

delve into *Project Drawdown: The Most Comprehensive Plan Ever Proposed to Reverse Global Warming* by Paul Hawken or the project's website, which is full of resources on climate actions that work at scale. My own recommendations were The *Future Earth: A Radical Vision for What's Possible in the Age of Warming* (an incredible, scientifically-informed piece of speculative fiction by meteorologist-turned-climate journalist Eric Holthaus, in which humanity does what is necessary to tackle climate change) and an episode of Rob Hopkin's podcast *From What If to What Next* in which he interviews economist Kate Raworth, author of *Doughnut Economics*, and Marieke van Doorninck, the Alderman for Sustainability and Urban Development for the City of Amsterdam, to discuss "What If Every City Used Doughnut Economics?"[25] Each one of these touches upon the built environment as part of a wider – technologically possible – societal transformation and integrates climate literacies and circularity beyond the scale and scope of an individual building.

Storytelling assignment

Instead of a traditional – albeit virtual – architecture school "Crit" format for our interim review, we invited our team to present their work as a five-minute story, complemented by whichever visual medium they felt most befitting to storyboard the narrative behind their explorations.

Complementing the students' assignment were some excellent examples of climate storytelling and speculative fiction:

- "A Day in 2030" – an animation based on a highlights reel compiled from *From What If to What Next*'s time-travelling to 2030 exercises
- "A Message From the Future With Alexandria Ocasio-Cortez" – an illustrated story that begins with early climate warnings pre-2000s in America and imagines the kind of transformations possible if the country boldly doubled-down on a Green New Deal in the wake of 2018's IPCC Special Report
- "A Message From the Future II: The Years of Repair" – a sequel envisioning where America could head if lessons were taken from the COVID-19 pandemic responses and synthesised into a Green New Deal
- "Prelude to a Great Regeneration" by Imagine the Future – a piece of illustrated speculative fiction that envisioned the winter of 2020 as the climate turning point that we needed at the time

Alongside storytelling examples, we recommended the *Building Sustainability Podcast* episode, where host Jeffrey Hart and his guest Rob Hopkins discuss "What if we imagined a better future?" – delving into the writing of *From What Is to What If*, Rob's permaculture teaching that resulted in Energy Descent Plans and his role in the founding of the Transition Network – a community climate action phenomenon, which has become an international movement. Beyond giving the students inspiration in different mediums to ensure their accessibility and tones of delivery, these storytelling examples and discussion were intended to build upon our brief's intention to envision futures in which climate adaptation leads to a community's thriving and to offer examples of what people are already capable of.

What the students presented was unlike anything we had ever seen from architectural education before, and not one of them had imagined an individual building being the result of their project. Simultaneously, two stories were told of repair, one in Northern Europe and the other in the South, of disassembling existing buildings and gathering wasted resources to create places for people to dwell and to use once more. Responses to the recurring theme of water management were envisioned very differently: one student opting to reimagine an industrial lot in Hackney as both a community garden and sponge as part of a new London-wide flood resilience infrastructure (Figure 4.3). Another student embraced seasonality and a return to more nomadic ways of life in response to seas rewriting the Belgian coastline – envisioning a timeline of homes becoming

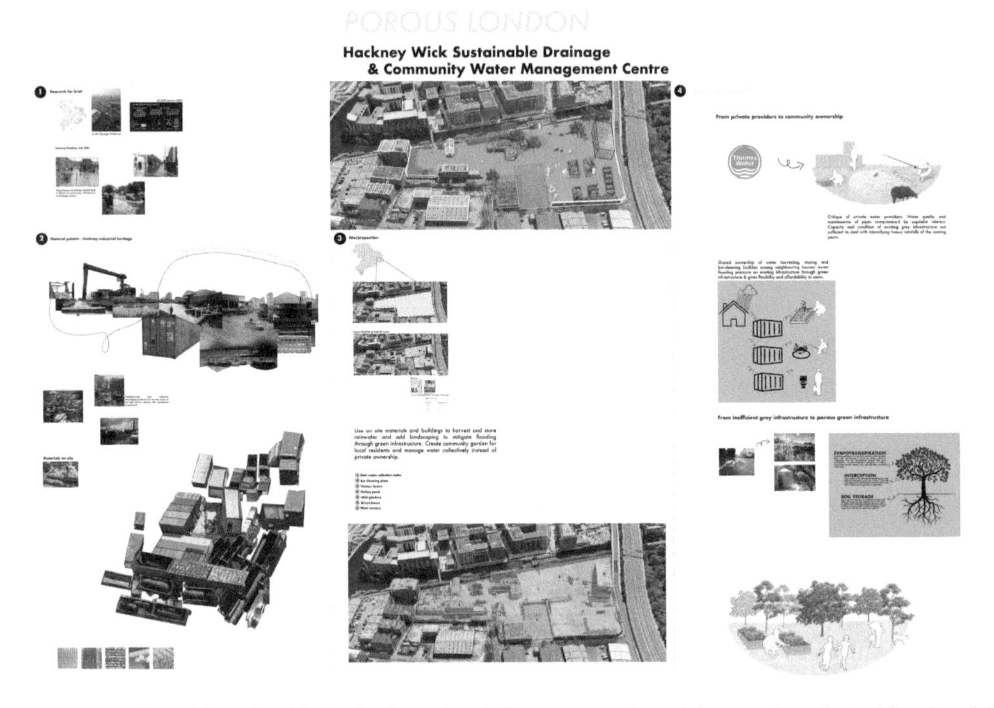

FIGURE 4.3 Imagining the kind of adaptation infrastructure that cities need, and stacking it with community uses and purposes

Source: Johanna Moro

FIGURE 4.4 Reimagining and deconstructing the redundant infrastructure of the fossil-fueled era

Source: Peter Monos

lighter, and more mobile, to avoid extreme conditions, whilst certain typologies of buildings remained static to support this such as schools, hospitals, and libraries of things. Symbolic, poetic prose accompanied an epic tale of a recently built oil company's headquarters being the final straw in Budapest (Figure 4.4), before being disassembled and redistributed to fulfill purposes that nourished the city – leaving a giant, structural skeleton behind. We were then offered a vision of China's abandoned railway carriages and stock, discarded before its time, being repurposed to send relief or entertainment where needed and to foster greater connection across distances. The seventh project took the form of a fire festival: a cultural intervention in response to an inhibiting policy, educating those in Portugal about the place of fire in traditional forest management, in which its infrastructure, pavilions, and sculptures are realised using salvaged materials that survived a prior wildfire, and its conclusion is a symbolic replanting and rewilding (Figure 4.5).

FIGURE 4.5 Culture and cultural activations of space as climate adaptation and educational infrastructure

Source: Daniela Martins

Each student's exploration imagined creatively re-using materials – in very practical ways and the regenerative stewardship of built and wild spaces, far beyond the boundaries and the time horizons of what one once considered the role of an architect or infrastructure. Without fail, every single one of them was more delightful, inspiring, exciting, and more fun than any design brief that I was issued during my own architectural education, and given more time than was available, these would have ended up as a truly visionary body of work, or likely – as Duncan Baker-Brown remarked after their final presentation to the wider cohort – a graphic novel.

Conclusions and replicability

What began as an invitation to imagine new uses for the buildings and infrastructure which would be rendered redundant by a 2 °C warmer world blossomed emergently into a replicable format of not only introducing circularity into architectural education but of simultaneously giving students an opportunity to deepen their climate literacies, autonomously and collaboratively; to look differently at their surroundings; and to give thought to possible futures. Flattening traditional studio hierarchies – inspired by our time in activist and campaigning spaces, especially the Architects Climate Action Network – was a key aspect of the studio's dynamism and made space for our students to learn from one another and work together of their own initiative, whilst we supported them and their learning. It also led to our tutorials – and even our interim review, our session of storytelling – being a lot of fun despite the usual heaviness of learning about climate literacies. As the climate crisis cascades violently into other crises all around us, exercises that make time for students to envision, design for, and to collectively illustrate possible futures in which communities have gone beyond adapting to a changing climate and are flourishing could become one of architectural education's greatest contributions.

This replicable, place-based, architectural, and infrastructural investigation and exercise in speculative fictions and the radical imagination could, in turn, be repurposed as a Week 0 exercise to ground design studios in the reality of a climate emergency and broaden students' notions of site analysis to include the infrastructure around us and to take stock of what materials are available in any given local area. Or it could be developed further and run as a lightly facilitated class or module. This would fuse and reconnect the cultural and the technical – introducing architecture students to the infrastructure of the Anthropocene, its impending redundancy, and its place alongside construction in working towards more circular economies – before inviting them – just as we did for the School of Re-Construction – to reimagine it for the unpredictable climate futures that are on the way. Should end-of-year exhibitions at schools of architecture begin giving space to such regenerative portals into the future on an annual basis and invite the world outside their walls to see them, it could prove transformational.

With the infrastructure of the Anthropocene approaching the end of its life and 80% of the buildings that shall be in use in 2050 standing up around us today,[26] the role and the work of the architect must radically change – from creation and conservation to stewardship, facilitation, and reimagination. This requires nothing short of a cultural transformation in architecture and construction, beginning and not ending in their education.

If we are to flourish despite the changing climate, we need to begin seeing interventions that go way beyond the building-at-a-time scale, such as societal-scale projects like decentralised, neighbourhood-by-neighbourhood retrofit plans, the phased deconstruction and relocation of materials from flooding settlements, and the decommission of infrastructure that we cannot save from the sea. The barriers to such actions are political – not technical – and perhaps the new role of the architecture school is to sow seeds for a Just Transition, empowering students to collectively imagine and illustrate what these technologically possible, hopeful, and irresistible futures could look like. After all, imagining, fighting for, and realising such futures as the climate breaks down is not simply a matter of climate and intergenerational justice but, increasingly, a matter of survival,[27] for those that we know and the billions we do not.

Raw 2 participants: Daniela Martins (Portugal); Johnanna Mono (England); Yuwei Ren (China); Katrien Devos (Belgium); Marvina Sinjari (Greece); Pauline Harou (Belgium); Peter Monos (Hungary).

References

1. Andrew Fanning, Dan O'Neill, Jason Hickel, Nicolas Roux, William Lamb, Julia Steinberger, Beth Stratford, Kate Raworth, and Katherine Trebeck. *A Good Life for All Within Planetary Boundaries*. University of Leeds, 2023. https://goodlife.leeds.ac.uk/
2. Daniel Christian Wahl. *Designing Regenerative Cultures*. Dorset: Triarchy, 2016. pp. 24–25.
3. Sophie K. Roas. *Radical Intimacy*. London: Pluto Press, 2023. pp. 117–118.
4. Joe Zadeh. "Concrete Built the Modern World. Now It's Destroying It." *Noema Magazine*, 2022. www.noemamag.com/concrete-built-the-modern-world-now-its-destroying-it/
5. Amanda Sturgeon. "Buildings Designed for Life." In *All We Can Save: Truth, Courage, and Solutions for the Climate Crisis*, edited by Ayana Elizabeth Johnson and Katherine K. Wilson. New York: Penguin Random House, 2020. p. 166.
6. David Wallace Wells. *The Uninhabitable Earth: A Story of the Future*. London: Penguin Audio: Audible, 2020.
7. International Energy Agency. "Global Electricity Demand Increases and Renewable Energy Generation Growth, 2019–2025." 2021. www.iea.org/data-and-statistics/charts/global-electricity-demand-increases-and-renewable-energy-generation-growth-2019-2025
8. Jason Hickel. "Quantifying National Responsibility for Climate Breakdown: An Equality-Based Attribution Approach for Carbon Dioxide Emissions in Excess of the Planetary Boundary." *The Lancet*, 2020. www.thelancet.com/journals/lanplh/article/PIIS2542-5196(20)30196-0/fulltext
9. Andrew L. Fanning and Jason Hickel. "Compensation for Atmospheric Appropriation." *Nature*, 2023. www.nature.com/articles/s41893-023-01130-8
10. Creative Denmark. "Designing the Irresistible Circular Society." *Creative Denmark*, 2021. https://ddc.dk/wp-content/uploads/2021/10/creative_denmark_white_paper_designing_the_irresistible_circular_society_small.pdf
11. International Energy Agency. "Net Zero by 2050: A Roadmap for the Global Energy Sector." 2021. www.iea.org/reports/net-zero-by-2050
12. Intergovernmental Panel on Climate Change. "Special Report Global Warming of 1.5°C." *IPCC*, 2018. www.ipcc.ch/sr15/
13. World Meteorological Organisation. "Atlas of Mortality and Economic Losses from Weather, Climate and Water-related Hazards." *WMO*, 2021. https://public.wmo.int/en/resources/atlas-of-mortality
14. Alejandra Borunda. "Weather Shows Evidence of Climate Change Every Single Day Since 2012." *National Geographic*, 2020. www.nationalgeographic.co.uk/environment-and-conservation/2020/01/weather-shows-evidence-of-climate-change-every-single-day-since-2012
15. The Guardian. "Uncharted Territory: UN Declares First Week of July World's Hottest Ever Recorded." *The Guardian*, 11 July 2023. www.theguardian.com/environment/2023/jul/11/uncharted-territory-un-declares-first-week-of-july-worlds-hottest-ever-recorded
16. Favianna Rodriguez. "Harnessing Cultural Power." In *All We Can Save: Truth, Courage, and Solutions for the Climate Crisis*, edited by Ayana Elizabeth Johnson and Katherine K. Wilson. New York: Penguin Random House, 2020. pp. 121–127.
17. Kendra Pierre-Louis. "Wakanda Doesn't Have Suburbs." In *All We Can Save: Truth, Courage, and Solutions for the Climate Crisis*, edited by Ayana Elizabeth Johnson and Katherine K. Wilson. New York: Penguin Random House, 2020. pp. 138–144.
18. Emily Elhacham, Liad Ben-Uri, Jonathan Grozovski, Yinon M. Bar-On, and Ron Milo. "Global Human-Made Mass Exceeds All Living Biomass." *Nature*, 2020. www.nature.com/articles/s41586-020-3010-5

19. Jason Hickel. *Less Is More: How Degrowth Will Save the World*. London: Heinemann, 2020. p. 156.
20. World Business Council for Sustainable Development. "Net-Zero Buildings: Where Do We Stand?" *WBCSD*, 2021. www.wbcsd.org/Programs/Cities-and-Mobility/Sustainable-Cities/Transforming-the-Built-Environment/Decarbonization/Resources/Net-zero-buildings-Where-do-we-stand
21. Emily Elhacham, Liad Ben-Uri, Jonathan Grozovski, Yinon M. Bar-On, and Ron Milo. "Global Human-Made Mass Exceeds All Living Biomass." *Nature*, 2020. www.nature.com/articles/s41586-020-3010-5
22. Caroline Hickman, Elizabeth Marks, Panu Pihkala, Susan Clayton, R. Eric Lewandowski, Elouise E. Mayall, et al. "Climate Anxiety in Children and Young People and Their Beliefs about Government Responses to Climate Change: A Global Survey." *The Lancet*, 2021. www.thelancet.com/journals/lanplh/article/PIIS2542-5196(21)00278-3/fulltext
23. United Nations. "The Paris Agreement: What Is the Paris Agreement?" https://unfccc.int/process-and-meetings/the-paris-agreement
24. Damien Gayle. "Millions Suffering in Deadly Pollution 'Sacrifice Zones', Warns UN Expert." *The Guardian*, 2022. www.theguardian.com/environment/2022/mar/10/millions-suffering-in-deadly-pollution-sacrifice-zones-warns-un-expert
25. Rob Hopkins. "From What If to What Next Episode Seven: What If Every City Used Doughnut Economics?" *Podcast*, 2020. www.robhopkins.net/2020/08/17/episode-seven-what-if-every-city-used-doughnut-economics/
26. World Green Building Council. "Bringing Embodied Carbon Upfront." *WGBC*, 2019. https://worldgbc.org/advancing-net-zero/embodied-carbon/
27. Alex Khasnabish and Max Haiven. "Introduction." In *What Moves Us: The Lives & Times of the Radical Imagination*, edited by Alex Khasnabish and Max Haiven. Nova Scotia and Manitoba: Fernwood Publishing, 2017. pp. 1–6.

5

RAW 3 'THE RE-USE IMAGINARY'

Graeme Brooker, Louis Destombes, and Hugo Topalov

FIGURE 5.1 Members of The Reuse Imaginary team screen-grabbed and then reimagined into the Ghent Altarpiece by Jan van Eyck (1432/2021): Student and staff team – (L to R) Graeme Brooker, Hugo Topalov, Kataryzyna Podhajska, Gabrielle Kawa, Wang Chung Cheng, Tess Hillan, Mariam Abuelsaoud, Timothy Danson.

DOI: 10.4324/9781032665559-10

This chapter has been made available under a CC-BY-NC-ND 4.0 license.

Introduction

If the early 21st century can be characterised as a time of resistance, in particular to the climate emergency, how can the built-environment and the interior respond to these ongoing challenges? The 'Re-Use Imaginary' is a workshop that explores the fact that in response to a world with finite resource, the very-near future of the built environment will be focussed solely on the re-designation of all existing matter. New-build and single-use processes will soon become obsoleted distinctions for making cities, buildings and interiors. Instead, re-using the existing site, with its matter already in situ, will provide all of the material for the appropriation of the existing to remake something anew. The Re-Use Imaginary workshop explored how particular domestic interior spaces could be transformed through the re-use of raw, useless and by-product materials, the hybrid, the offcut and the wasted and the discarded, via their appropriation in a manner that colonised existing buildings in which the participants of the workshop were residing.

This was a workshop which had two objectives: to make re-use processes and values explicit and to do so by emphasising the participants' awareness of the abundance of material at their fingertips in their own domestic spaces. The domestic interior was deliberately chosen as the location for the project because it addressed the recurring perceptions about interiors that prioritise the misconception that this discipline is focussed upon the decoration of luxury homes. On the contrary, the subject of the interior has been at the forefront of reworking existing buildings since early civilisation. Re-use and interior architecture/design/decoration are synonymous with the adaptation of what is already in existence. It is the view of the agents in this discipline that buildings are conceived and constructed primarily in order to host interior spaces.

Less provocatively, the workshop was undertaken during the global coronavirus pandemic and, therefore, we were all 'locked-up' in our various 'lockdowns', affording us the time to not only reflect on our domestic situation but also to carefully reconsider our relations with our cohabitees and communities and to renegotiate our own sense of what was public and private. Thus, everyone's home became not just our place of work and leisure but the focus of our contemplations on the things surrounding us. The imaginary of domesticity and its re-use was, therefore, deemed as the most apposite vehicle for this workshop in order to emphasise how our home-environment could be viewed as a stockpile of available material.

Re-Use

For the benefit of participants, it was very useful to establish what was meant by the terms being used. Re-use is a phrase utilised to describe the repeated application of an item to different contexts, from which it originates, time again and time again. One of the clearest descriptions I have enjoyed is by Hegewald and Mitra. They suggest *re-use* can be applied to anything – it might utilise

> an object, an edifice, building materials, a style, a law, a concept, a form of governance, an idea or anything else. It is a deliberate and selective process in which existing elements are borrowed and taken out of their former surroundings to be applied to a fresh context.[1]

We utilised this clear and fundamental explanation in order to emphasise the *locality* of re-use, that is, how things around us can be repurposed. This was undertaken with the subtext of the advocacy of a social and ecological spatial well-being attained through reducing excessive forms of consumption. In this workshop, *re-use* was a term used to open up discussions and to frame how your home may be viewed as a stockpile of resources to be repurposed, adapted and never discarded. This discussion was introduced in order to open up responses to the limits of forms of growth and, in particular, the global economics of capitalism. We read parts of degrowth texts together, such as Kallis (2018), Hickel (2020) and texts on repair and maintenance, such as Mattern (2018), Jackson (2013) and Sample (2016),[2] in order to substantiate our aims of exploring the prioritising of care, equitability and the common values of the local – all critical elements of re-use. We wanted to outline resistance to all forms of extractivist environmental degradation, unsustainable resource depletion and excessive and wasteful processes of built-environment developments. In short, we emphasised approaches that foregrounded maintenance and the prioritisation of the incorporation of the repair of the existing. We advocated for the exploration of the care of space and its occupants, the incorporation of entropy and decay, do-it-yourself approaches such as salvage, hacking, and so on. This workshop had ambitious and bold aims, and we hoped that in some way, we, the convenors of the workshop and its participants, at least initiated a conversation and planted some seeds for the students responses to their own domestic environments and what it has and could mean during the COVID pandemic era.

Imaginary

The other important term – *imaginary* – relates to the existence of something unreal or yet to materialise. I decided to use this term to counterpoint the *material* dimensions of re-use with a few tangible dimensions. *Domestic* was added in order to reinforce the link to the participants' own situations along with combining an ambition to signal how the workshop could unfold through rethinking domesticities. Home is a complicated space:

> Once a sanctuary from prying eyes, the home is now a geotagged broadcasting studio from where we share our most intimate moments and display our carefully curated online identities.[3]

In the last two years, as we have resided in our domestic interiors in ways which we have never previously considered, almost overnight, our homes became our workplaces, our health centres and the location of our entertainment. They became the places where we had to form a bubble, to shelter or even isolate from other inhabitants both within and outside of them. They also became more public than ever as we all examined each other's backdrops in the Zoom calls that we made to each other. The front doorstep, the window and the balcony became the links to the outside, where we touched hands through glass and sang and clapped to each other in order to show our appreciation of both health workers and the delivery people keeping us supplied with goods and food. The intensity of our focus on our interiors has arguably not only increased our desire to alter them in order to satisfy these new needs but, more so, stimulated our quest to find out more. An imaginary-based approach offered the vehicle for these reflections. All of these conditions and situations are not new. The interior has always been public. It has always contained a close connection to our place of work, been a hub for entertainment and the focus of our health. Through 'imagineering', this workshop set out to chart and document some of these new explorations and to explore how the fundamental purposes of containing human inhabitation, itself a fluid and everchanging slippery and indefinite entity, might evolve and have changed and be projected into the future.

Even before the pandemic, humans spent 90% of their existence indoors.[4] That is all but half a day a week *inside*. Interiors do not just provide shelter (progressively so in an air-polluted, COVID-restricted and climatically challenged time), but they also actively shape how we work, learn, love, travel and take our leisure. All of our lives take place in a wide range of inside environments, yet it is rare for their inhabitants to profess a deep understanding of the spaces that they are within. The ambition for the domestic imaginary was to give insight into both re-use and domesticities – through the reappraisal of the familiar or that which has surrounded us for quite a while as well as the revealing of how what is a recognisable and familiar yet profoundly enigmatic spatial entity can be full of both physical and immaterial raw material for re-use.

In the early stages of the imaginary, the participants agreed that home was a universal theme. It described a place, often a collection of rooms, which encapsulated all forms of human relations and behaviours. It also is the space that personifies experimentation in how we live, speculations on finance, ownership and the *asset*: the commodifying of privacy. The home is the site for improvement, maintenance, regulatory governance and standards. And, of course, in current times, it is the place where we all now undertake work. In essence, the home has always been the most public of private spaces. This agreement set the scene for the participants of this workshop and really focussed their reflections on the speculations on the home that had surrounded them that year.

The workshop took place in three phases over the two weeks;

- **Imaginary** – abstractions of the essences of home through imagery and modelling
- **Inventory** – indexing, cataloguing the room and applying the imaginary to it through utilising a developed surface method
- **Super-Composite** – reconstructing all re-used imaginary through the amalgam of the raw material of the participants immediate environment and the results of their learning so far

Technique: collage

To note, the chosen medium for the workshop was a hybrid digital/manual collage approach, asking participants to deploy their selection and editing skills, with both scalpel and laptop. In general, when teaching, my students are usually required to illustrate their thoughts by editing, judging, selecting and translating, then sticking stuff together in order to unlock all kinds of systems of signs, imagery, words that are then combined to make something with completely new meanings. Collage is used because each fragment never quite shakes off the load of its former self, be it an image culled from a magazine, a book, online or on a screen and so on. Collage is a great re-use educator. The joy of collage is the careful compacting of meaning upon meaning to form an amalgam, a composite of references, manipulated into relaying something new. Formal composing structure and poetry are also a part of the mix. In the hands of the experts, they can be funny, knowing, amusing and also unusual and striking. Collage processes provide an education in judgment and editing, just like the spaces my students are asked to make judgements and edits of to re-use in their work. Therefore, as you will see, collage was the chosen medium for the project.

Phase 1: imaginary

With its etymological roots in *image,* the imaginary describes the formulation of something that is realised through its inexpressible or imaginative qualities. In order to 'ground' this slippery quality, each phase of the workshop started with the examination of a precedent. In this first phase, we examined Tim Peeters' Uyttenhove House. This was a project that converted a generic suburban house in Sint-Amendsberg, a suburb of Ghent, Belgium. It was of interest to us as we could see how some of the traditional traits of housing were still represented in the project, yet its re-use had entailed the unusual enlargement of its footprint, literally through swallowing the house whole with a new 'wrap' of a building. Peeters describes how the old brick façade became a souvenir of the existing home – a child-like portrayal of the family home. At its most simple explanation, the *imaginary* described representations of home, often the material representations of interior space and the lives that occupy them as told through images in photographs and film and through novels and storytelling.

FIGURE 5.2 A collection of some of the choices of paintings. Students all chose their own images to bring to the workshop, a revealing diagnostic process.

After we analysed and discussed the precedent, the workshop commenced with each participant selecting a series of paintings, from the 15th century onwards, that depicted domestic settings. (Figure 5.2) Each prioritised the raw, the useless, the by-product, the hybrid and the offcut. We focussed the imaginary into utilising representations of domestic spaces in painting only, and it was stated explicitly at the beginning of the project that the raw material must take the form only of paintings of domestic spaces from the 15th century onwards. Why? The date was unimportant, but the medium was critical. Paintings of domestic spaces contain the explicit translation of a number of elements of interiors which other mediums do not necessarily articulate. For instance, whilst a photograph or a digital rendering of a domestic interior capture people, light, objects, spaces in particular ways, I chose painting as it reinforced particular elements through their choice and positioning and their articulation in paint rather than any other medium. Paintings are constructed images that are inherently fictitious. Of course, they often depict real or actual spaces, objects and people, but it is the emphasis on technique, composition, medium and style, when depicting the elements from which they are composed, that arguably ensure that they increase or emphasise their staging or reconstructive dimensions. Mediums such as photography or film also contain the potential to render their subject fictitious, but arguably, the connotation of reality that the mediums' relations with its subject inherently contains ensures that they have the capacity to maintain some semblance of reality. I am fully aware that this is a contentious observation and one that requires greater unpacking, and in the workshop, we debated this approach, but we felt it was the way in which to proceed. In order to distil the essences of domesticity, my students self-selected 15 images of paintings from the 15th century onwards, each representing home. The paintings chosen, as you can see in Figure 5.2, curiously very Western-centric, became poignant reflections and indicators of what each of the participants thought was important in their home, but they also provided revealing indications of their own state of mind. This was particularly pertinent when images were chosen with people placed into them. It was as though they represented the student's own situation. Some were singular, forlornly staring at spaces outside of the frame of the image (Hopper), waiting for something to happen. Others were a part of a group, socialising, a restricted pleasure at the time of the workshop.

Fundamentally, recurring themes that emerged from this process were light, people and their position in space on their own and in relation to others, framing, view, furniture and objects. Further analysis revealed choices made through participants' heritages, cultural references and intuition but, more so, careful thoughts as to how the spaces and the compositions spoke to each of the students. Possessions and objects featured heavily with each image expressing as much of what was inside the frame of the image as much as what was left out. As a diagnostic process, it provided its own revealing portrait.

After the choices were arranged in a Miro board, overlaps and same choices edited and the key themes discerned, the second part of the imaginary phase was what was described as *the disassembly composite*. This was an approach which required each participant to unpick each image, selecting key elements, in order to form a super-composite of the 15 images. This required the students to compact all of them together as one super-collage of a home. It was not only an attempt to find a way in to the project, it was also an ice-breaker for the students and me. These *imaginaries* offered the beginnings of the translation of the components of home futures, laid bare through their editing, removal and recontextualising with scalpel and glue. The resultant collages were startling. From the raw material of the paintings, the students carefully built-up super-composites.

To guide us through this part of the process, we explored Ben Nicholson's compelling collages from Appliance House, a project where he neatly dissected images of domestic and consumer goods, hence the word *appliance*, from magazines and catalogues in order to form collages or the plans and sections of a house for the *Kleptoman*. He was the mythical Figure or occupant of the space, who, from the title Nicholson gave him, was obviously disposed to acquire material with which to build his home from scrap and detritus. The dis-assembly collages were beautiful (Figure 5.3). Some were artfully arranged to draw out the essences of form,

FIGURE 5.3 The dis-assembly composites

space, light, others responded more to the formal constraints of painting such as the triptych or using frame to discern space. Each dis-assembly composite gave the participants a base from which the translation of a set of house spaces had taken place into a new form of interpretation. Themes of framing and light predominated. Others of people and furniture, at the centre of each composition were rooms, each the focussed interior of human occupancy. These composites provided the base or starting point from the conflation of the students' own houses with the project was beginning to take shape. From each image, we could begin to discern emerging essences of domestication that each student wanted to draw out and to explore. The next phase was to make this connection more explicit.

The third part of the imaginary phase was for the participants to make the leap from the translation of the paintings, through the disassembly process and their reinstatement and re-use into the new context, into three dimensions. In a physical setting, we would normally undertake this process with card and model-making. This was still an option, and one of the students took this process on but with some difficulty due to accessing materials and the space in which to make with card and glue. Most of the students worked digitally to make their composite imagery undergo a third interpretation – into actual rooms.

For guidance in this phase, we consulted the work of Gordon Matta-Clark and the way that he would formulate space through the removal of elements and parts of buildings, in contrast to the normal processes of the addition of elements and objects. In essence, the art of clarifying through subtraction was important to understand. In this third phase, participants were asked to consider how to reinforce and accentuate the spatial features and elements of their earlier sequences of editing and recomposition and to extract, through re-use, the essences of their domestic imaginaries. This was in order to demonstrate how these phases could, in a relatively abstract manner, and also whether consciously or not so, assimilate the essences of domesticity. It also brought into play the idea that surrounding each person were the raw materials for re-use. The resultant imaginaries provided a rich and diverse set of rooms and homes. Spatially complex and unusually so, in many cases, the 'raw' imaginaries provided new and compelling composites of domesticities. The next sequence of the imaginary was to carefully embed these works and thinking into the participants' own domestic situations. This would be done through processes of *inventorising*.

Phase 2: the inventory

> *Inventarium . . . list of what is found . . . from Latin Inventus, past participle of invenire, find, to come upon.*[5]

To make an explicit link between the imaginary and the participants' domestic situations, the group initiated what I would call an inventories-based approach. The term *inventories* is a word borrowed from 15th century practices of the recording of lists of primarily domestic goods, usually with their estimated values, and often used to surmise the belongings and land of someone deceased. These were then used to assist in the distribution of the deceased's estate. Essentially, *inventory* describes the processes of listing collections of the family's belongings for their subsequent sale or redistribution. I argue that inventory-based approaches foregrounds practice-led or hands-on research like this project, primarily because it prioritises auditing and valuing, and these are the methods of collating, utilising and implementing the materials that have been found on a site. The found or indexed elements, such as the possessions in a person's room or home, can be taxonomized, catalogued and readied for re-using. Instead of the application of a preconceived framework of enquiry, participants explored the existing, or the found, matter on their 'site', in order to then deduce and formulate approaches to its subsequent redesignation and re-use.

To guide us in this approach, we used Sarah Wigglesworth's classic drawing of the dining table in her studio. It was a drawing that demonstrated, on one hand, the idea of designing a building through understanding the life of its inhabitants and, on the other, illustrated the intentions of a project's inception, ideal and messy realities. I tasked each designer with inventorising the room in which they were undertaking the project. Some were methodical, mixing image with lists (Figure 5.4)

Others less so, preferring to rely on photographs and lists of their own rooms.

In order to move from the sometimes fixed or objective qualities of the inventories, the second part of this phase of the workshop involved the making of a developed surface drawing of the room and the inventorised objects within them. We were guided by the work of Robin Evans, particularly in his seminal *Translations From Drawing to Building and Other Essays* and the essay 'The Developed Surface'. In the essay, Evans states:

FIGURE 5.4 The inventories

> The developed surface representation obliterates the connection between an interior and its surroundings. With its exclusiveness accentuated, an interior so drawn can flourish on its own identity and need receive none of its attributes from its relationship to anything that impinges on it from outside.[6]

This was an important device to utilise because we wanted the designers to reclarify their own rooms whilst defining their contents and the material available to be re-used. 'Outside' was of little importance. The Developed Surface was a technique that could be both objective, a listing of the contents of the room, and a translation. Its reduction of external influences was useful and reinforced the internal nature of the work being undertaken. In this part of the process, some students emphasised the objectivity of the inventories Developed Surface approach (Figure 5.5)

Others became more expressive and fully utilised the collagists 'licence' to extract and loosen an image from its moorings in order to float it into another context. This process gave the creators a really important insight into their own possessions and goods – inventorying their own immediate context and embedding it into the room as though a catalogue of elements for (re)use.

The Developed Surface images fully interiorised the spaces, and when at their most successful, they fully amalgamated possessions and rooms into a whole. It was at this point that we entered the final phases of the project and that the work could start to be brought together.

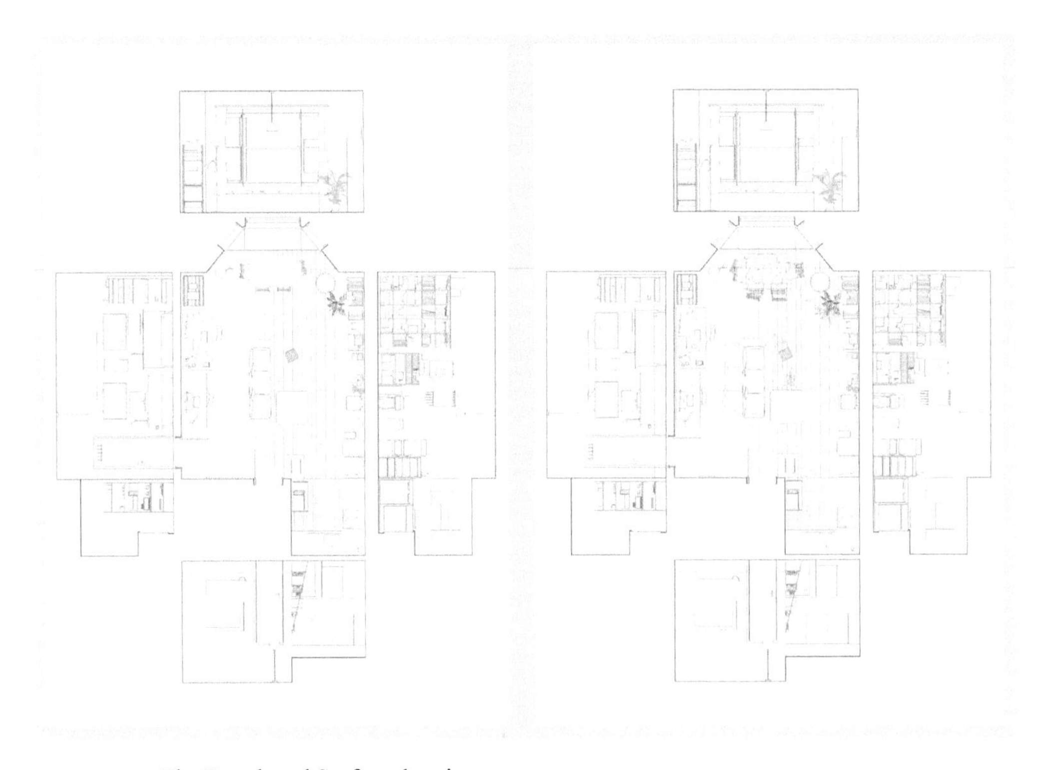

FIGURE 5.5 The Developed Surface drawings

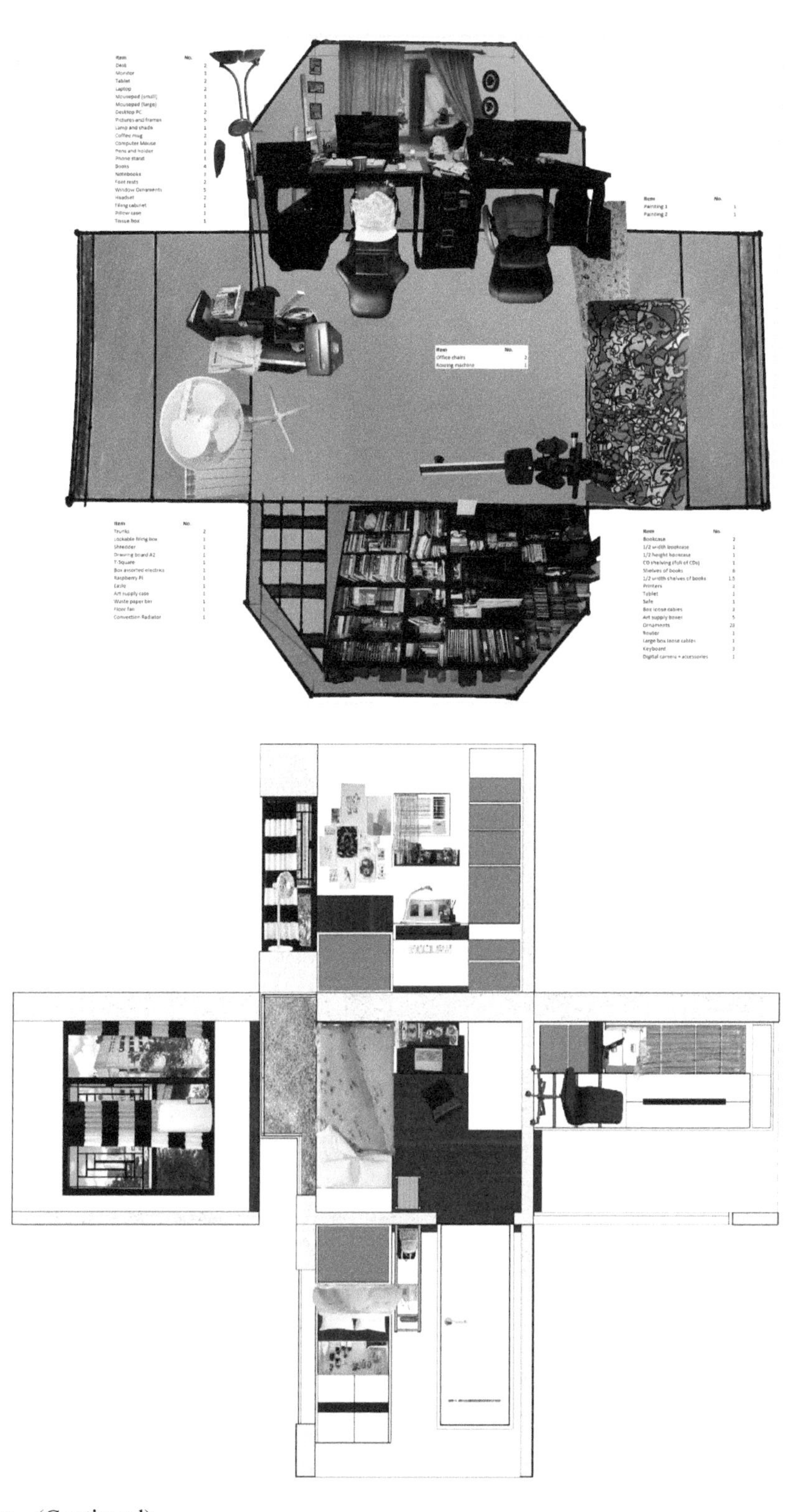

FIGURE 5.5 (Continued)

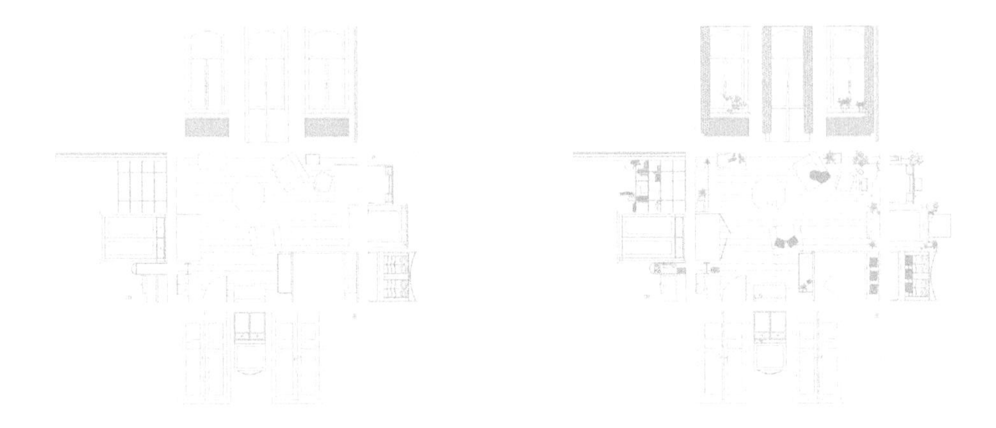

FIGURE 5.5 (Continued)

Phase 3: super-composite

The final phase of the project was where I wanted the students to recap all of the stages and to be able to present a coherent imaginary which had chronicled not just their project work but also the journey we had been on together. A composite is, of course, a compaction, a compression of layers, or the amalgam of the fusion of skin and structure. Here, we used it as a way of describing how all of the sequences of the journey could be compacted into one distinct image, model or output. There was no direct brief for this. At this point, it was totally unscripted and over to the participants to see what they could come up with. It was probably the most difficult part of the project because in the previous phases, the work and the processes were clearly set out. This phase was reliant on the successes of the earlier phases and what had already been produced, so its qualities were informed by what had gone on before, building upon each successive layer of previous work. (Figure 5.6)

The super-composites ranged from the addition of all of the drawings into one super-section, the fabrication of a building expressing the forms and features of the previous imagery and two highly decorative Developed Surface composites.

Each wrestled with the various components of the learning journey, compacting objects, elements, people into their imaginaries. I was very excited to see how the authors of the works were enjoying a relatively free and intuitive approach to the imagery, layering in their possessions, the elements of the paintings from before, into one whole amalgam. There were some surprising images where people were layered into the imaginaries. Others focussed on spatial composition and architectonic or formal qualities. What was evident from all of the participants' work was how they could understand how their current and personal environments could be realised as inherently complex combinations of not just material goods, possessions, rooms, spaces, but also intangible matter such as memories and emotions, all available for re-use.

FIGURE 5.6 The super-composites

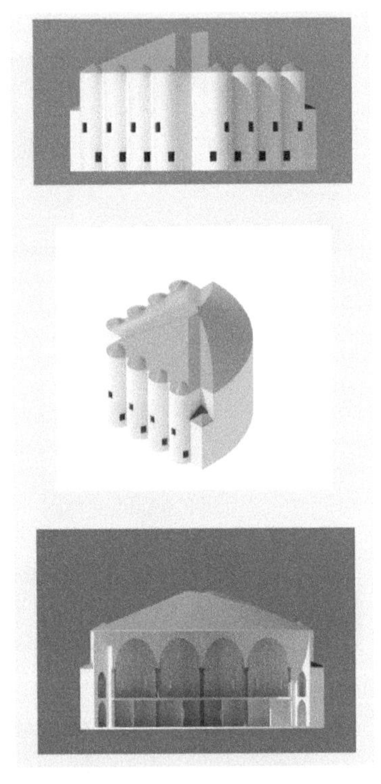

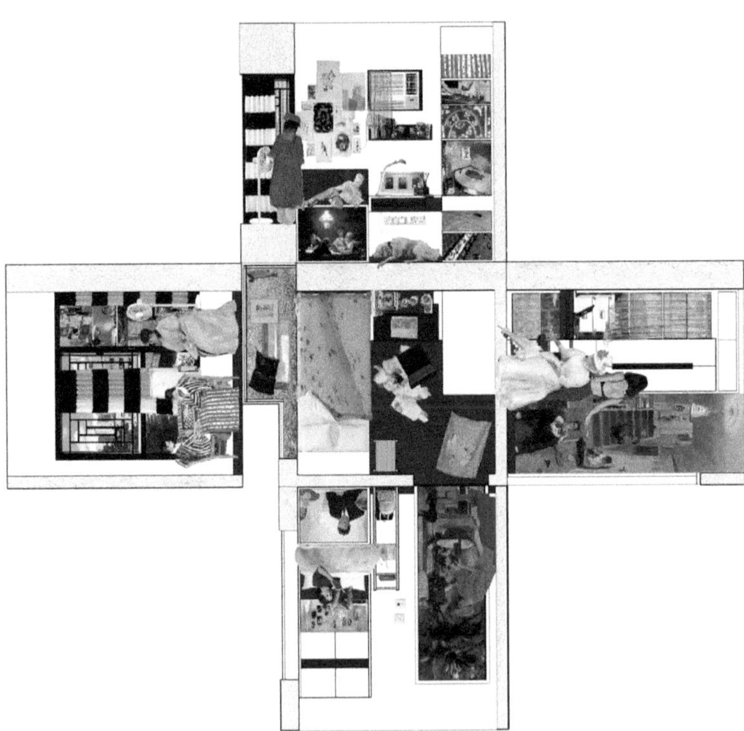

FIGURE 5.6 (Continued)

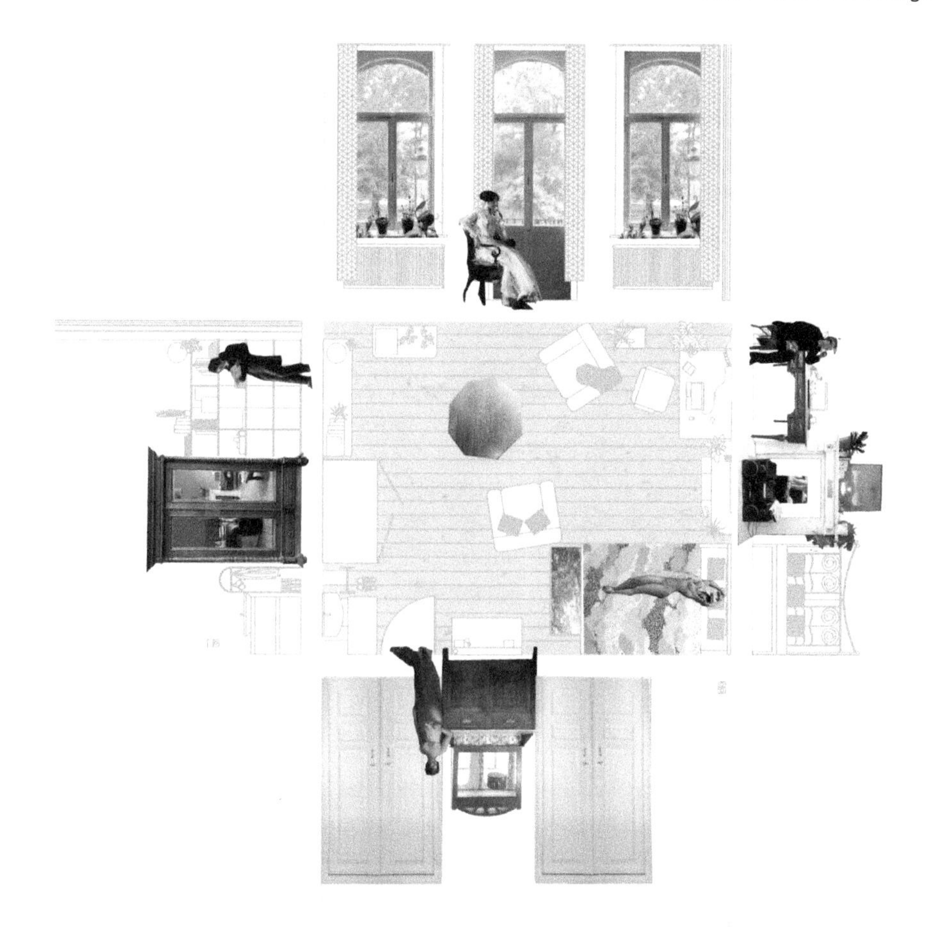

FIGURE 5.6 (Continued)

Endings

I opened this essay with a collage of the infamous painting by Jan van Eyck which the students had reordered by collaging each of us into the panels of the work. Done in their spare time, the collage summed up the spirit and comraderies of the group. We had fun, or it seemed that way from the various Zooms we made throughout the two weeks of the workshop. This collage spoke more meaningfully of the participants response to the ideas of the project, the workshop and the leaders of the group. There was a great sense of spirit, openness, care and fun in this project. It was not only a pleasure to convene and run, but it gave us all regular and meaningful contact with each other through the work we were doing in what was a difficult time in everyone's lives during the pandemic.

Because of this spirit, instead of the traditional conclusions or summing up of a project, I would like to emphasise the circular nature of the work in that its ambition was to open up new processes and future thinking for each of the participants, especially in relation to how it might feed into their own practices. In light of this, over a year later, I decided to contact the participants and ask them to reflect on the workshop and if it had had any impact on their work and thinking. It seemed fitting to draw this essay to a close with the words of the group. This was written by Tim Danson, an architectural technologist on behalf of the group:

> As an Architectural Technologist I'm interested in how things are put together and making things buildable so what better way than spending a summer school imagining it being all taken apart and then challenged to do something else with it? My time at the School of Re-Construction significantly redirected my studies and career designing buildings through the thought processes and techniques imparted. I was pushed to think hard about what was around me – was it raw, refined, assembled, glued or bolted and what 'could' happen to it next. I see just how much more flexible we can be with a bit of understanding, practice, good systems around us and the willingness to make the effort. Mining the Anthropocene has become the principal that I took away from the School of Re-Construction and which I now use to inform my design decisions – we don't need new materials; we need to use what we already have and this is achievable. We also need to think several steps further and make sure that it can be dismantled again in a way that enables multiple reuses or alternative uses. Just this week I was shown a photo of roof timbers still in use, dendrochronology dated as almost a millennium old and in good enough condition for another if detailed and maintained properly![7]

From Gabrielle:

> The Summer School of Re-Construction was for me a fun way to conclude my master's in Architectural Engineering from the Vrije Universiteit Brussel and my master thesis on the reuse of building materials. Since I, myself, worked on developing a guide for the integration of reclaimed wood and brick in architecture, I was interested to see and discover other but related practices on the same topic. I enjoyed the range of interesting keynote presentation, and the inspiring work others are doing to facilitate the reuse of building materials and components. Moreover, I liked how our workshop was different from the others, taking a more creative and artistic approach to the same challenging topic.
>
> During our workshop 'The Domestic Imaginary', we explored the concept of ownership through examining paintings and illustrations, how to make and visualise inventories in a creative way and studied through model-making the role and place of reclaimed elements in our own work. Through the interesting discussion and presentation sessions in our international group of students, with all a different background, we came up with a common perspective of how materials and elements can be reused in an explorative and creative way to achieve something new.
>
> Through the feedback from the other students in our group and the facilitators, Graeme and Hugo, from Bellastock, I got new insights in the overall topic of working with reclaimed elements, but also in my own work that I presented. They saw sometimes more in what I, myself, saw in my work. This really boosted my motivation and inspiration.
>
> The only negative thing to say about the workshop and summer school is that it is a pity that it could have been organised in real life. I think interacting with each other in person, and also having the informal moments, would have been very beneficial for both brainstorming as well as connecting our ideas on the Domestic Imaginary.[8]

Notes

1 Hegewald, Julia A. B. & Mitra, Subrata K. *Re-Use the Art and Politics of Integration and Anxiety*. Sage Publications, 2012. P3.
2 Kallis, Giorgios. *Degrowth*. Agenda, 2018. Hickel, Jason. *Lees Is More: How Degrowth Will Save the World*. William Heinemann, 2020. Jackson, Steven J. *Rethinking Repair*. MIT Press, 2013. Mattern, Shannon. *Maintenance and Care*. The Journal of Public Scholarship on Architecture, Landscape and Urbanism, November 2018. Sample, Hilary. *Maintenance Architecture*. MIT Press, 2016.
3 SQM: The Quantified Home. *Space Caviar*. Lars Muller, 2014. P23.
4 In May 2021, the Environment Protection Agency (EPA) stated that 93% of Americans spend their time indoors – 87% in buildings, 6% in cars. For Europeans, it is 90% indoors. The global average is between 90–98%.
5 Barnhart, Robert. *Dictionary of Etymology*. Chambers Publishing, 1988. P542.
6 Evans, Robin. The Developed Surface: An Enquiry into the Brief Life of an Eighteenth-Century Drawing Technique. In Evans, Robin (ed.). *Translations from Drawing to Building and Other Essays*. AA Files, 1997. P208.
7 Tim Danson via e-mail 7 November 2022.
8 Gabrielle Gawa via e-mail 25 November 2022.

6

USELESS 1 'DEEP RE-USE'

Jonny Pugh and Eddie Blake

FIGURE 6.1 *3D thing*, (day 7) – experimental 3D model using photogrammetric scans of brick fragments prepared for reuse.

Source: Derk Ringers at the School of Re-Construction, 2021

DOI: 10.4324/9781032665559-11
This chapter has been made available under a CC-BY-NC-ND 4.0 license.

When you close your eyes, it's easy to picture an old brick – its shape, dimensions, colour perhaps. You could even have a sense of its weight. We carry one around with us at all times – a virtual, generic one, which is deployed every time we think 'brick'. In reality, every brick is different, with its own history and idiosyncrasy – the closer you look, the more is revealed to you. When we design and build architecture, the brick is often treated as if it was the generic virtual brick you can picture rather than the specific real-world brick.

What does architecture look like when professionals and the public start to intimately understand the objects that make up their buildings? How does architecture change when its constituent parts are treated as specific objects rather than generic commodities? What implications might this specificity have on the culture of re-use in the realm of DIY making? By re-evaluating materials commonly considered obsolete, what new aesthetic tastes emerge?

We explored these questions through an experimental two-week workshop titled 'New Forms of Measurement' at the School of Re-Construction. The first week, 'Experimental Inventories', looked at tools that might capture a greater variety of material qualities. The second week, 'Montages', made use of the readings made by the new tools. From this process emerged the concepts of 'Deep Re-use'.

Context

Deep Re-use is an approach that responds to the ecological imperative to use fewer raw materials – focussing on the cultural case, alongside economic and technical aspects of re-use.

The construction industry consumes approximately 50% of all natural resource extraction worldwide.[1] A precondition for this wasteful practice is that end users are distanced from the materials that make up their built environment. A comparison with how consumption and waste are reported in fashion, products, or the food industry illustrates how this reality scarcely impacts popular imagination. Taking food as an example, the industrialization of production has created counter movements pushing for stronger critical public awareness, "empowering consumers to make healthy and sustainable choices".[2] This has been pushed, in part, by legislation of obligatory labelling requirements for ingredient provenance. In contrast, information about materials in construction processes is often accessible exclusively to professionals – end users and the general public have little contact with this information and are, therefore, often unaware of the potential environmental and cultural benefits from re-using the existing material world around them.

Despite the extreme urgency to transform the construction industry from a linear to a circular economy, we still lack viable mechanisms to properly value re-used materials. Commodity price is a useful, if inexact, mechanism that indicates the relative scarcity of a commodity in a market. However, that price does not account for carbon cost. It is the commodity price which most often dictates professional and non-professional material selection. So while a current lack of radical policy renders the re-used material market financially prohibitive in many regions, making different choices is hard for consumers. The intrinsic value of construction components is negligible in comparison with potential real estate profits, which incentivise demolition instead of deconstruction. *Deep Re-use* builds on the growing set of tactics which encourage re-use, for example, material passports.[3] However, rather than focussing on technical standardisation, *Deep Re-use* favours the additional transformative power of seeing objects through a cultural lens.

FIGURE 6.2 *Separation of bricks using non-specialist tools,* physical process experiment (days 2–5). Taking a lead from Ivan Illich's philosophy described in *Tools for Conviviality*,[6] Derk Ringers responds to the challenge of manual disassembly of brickwork: "*Bricks separated: 6/Amount of cleaning vinegar: 4 litres/Days soaking in vinegar: 3/Estimation of labour: 1.5 hrs*".

Source: Derk Ringers at the School of Re-construction, 2021.

To return to the ubiquitous brick – in the UK, "an estimated 2.5 billion bricks arise as demolition waste each year (almost equal to the number we use new each year), but only 5% are reclaimed for re-use, with the rest crushed for fill".[4] The prevalence of cement-based mortars in contemporary bricklaying makes effective brick deconstruction challenging. Yet a re-balancing of the labour market[5] and changes in popular aesthetic preferences could bring radical opportunities for job creation and new types of participation in making through re-use. The experiment begins with new ways of seeing.

The paradox of novelty

While most of the construction sector in the developed world thrives on an appetite for the 'new', there is an existing, if narrow, contemporary re-use industry. It is often focussed on 'architectural antiques' and or finishes that are considered 'high quality'. In the case of the former, the taste for antiques paradoxically relies on the exclusivity and scarcity of those materials. In the case of 'quality', this is often underpinned by ideas of durability based on a preconceived idea of single use. In both cases, this perception obstructs the potential for a radical re-used material revolution.

Deep Re-use has the potential to liberate us from such fickle perceptions of heritage and fashion and a limited imagination for alternative further lives of materials. As Cullen and Allwood

FIGURE 6.3 Collated frame extracts from the film *44 Doors and 35 Windows for the New Sala Beckett*, 2016. Carpentry elements of an abandoned Cooperative in Barcelona are represented in a film as characters in the story of their reinterpretation and re-use. Despite having no formal 'heritage' value, these doors and windows represented the time and love of a collective process, built by the cooperative members themselves in their free time, each making different decisions about the technique and design as they went along.

Source: Flores & Prats Architects

observe: "We rarely demolish buildings because their performance has declined, but because their value to owners or occupants declined, so they become unsuitable or undesirable".[7] The challenge lies in commodification.

Deep Re-use can be seen as an attempt to create bubbles of decommodification – places where different values re-emerge. Prior to being commodified, objects have specific individual use value. After becoming a commodity, that same object has a different value: the amount it can be exchanged for. This idea returns us to the old brick. The brick has a specific market value which you could look up right now, but you know it's not truly the same as the exchangeable brick. According to Karl Marx, this new value of the commodity is derived from the time taken to produce the good. When an object is commodified, all other considerations are obsolete, including morality or environmental impact. Marx went as far as claiming that everything would eventually be commodified: "The things which until then had been communicated, but never exchanged, given, but never sold, acquired, but never bought – virtue, love, conscience – all at last enter into commerce".[8] We may have got close to this reality, but *Deep Re-use* offers a potential way of stepping back from the precipice. Marx notes the danger of commodification being commodity fetishism and alienation.[9] The process of *Deep Re-use* de-alienates. It brings people directly back into contact with the real substance and use value of objects.

The *Deep Re-use* approach – an outline

Key to the approach of *Deep Re-use* is the 'measurement' of material. We start by asking some basic questions like: What is being measured? Who participates in and influences the measurement? Who has access to the measurements?

In response, the technique of 'Experimental Inventories' was developed, using standardised measurement methods (e.g., taking values for dimensions, optical qualities, structural or thermal performance), then integrating additional information borrowed from conservation methodologies (e.g., histories of origin, maker, popularity), and subjective readings (e.g., referential description of the form, moral judgement – is it the best of its type? – or a memory that is associated with the object). It is the historical and subjective readings that start to build up a cultural understanding – "immaterial values" as described by workshop participant Julia Flaszynska such as the "validation of craftmanship and emotional bonds".

Implicitly, *Deep Re-use* is a critique of common forms of material passports and points to ways that they could become participative and reflect a broader set of values. Even the format of recording is expanded from the use of a basic spreadsheet towards a broader set of

FIGURE 6.4A (left): *Brick selection AR* (still frame from animation, day 5). Derk Ringers speculates on a future model of an open-source material investigation app that could identify and share multiple levels of information for materials with re-use potential. Here, Ringers could be seen as a contemporary version of a 'prospector' (the mineral/mining detectors of the Renaissance era now suited towards mining of the Anthropocene in the 21st century), with a focus on new tools for observation.[10]

Source: Derk Ringers at the School of Re-Construction, 2021

FIGURE 6.4B (right): *Misuse montage* (1 hr exercise, day 6). Quick intuitive exercise exploring re-use languages using a pool of materials shared between students.

Source: Thomas Parker at the School of Re-Construction, 2021

'tools' including interviews, analytical drawings, photographic records, physical samples and 3D scans. The workshop pushed towards creating accessible, editable and active tools that allow for constant reinterpretation of the past and present.

The workshop's second phase, 'Montage', involved synthesis of these measurements. Compositional techniques were tested using the prompts: 'order', 'tesselate' and 'symmetry'. A variety of design and non-design approaches were adopted involving deliberate mis-use and re-contextualizations. One of the key design tools available to the *Deep Re-use* designer is placing re-used building materials or objects in dissonant contexts, or places that amplify or contradict their original use.

The *Deep Re-use* approach – tactics

Deeper readings might help us overcome negative associations with 'secondhand' materiality, in search of popular and radical aesthetic languages for re-use. For example, it is the proven provenance of the antique, that is, a deeper understanding of its cultural value, that really increases its exchange value. Looking at some exemplary student experiments from the workshops, we find some applications that come out of these 'New Forms of Measurement'.

In his investigation of bricks in the Netherlands, **Derk Ringers** looked at the challenges of disassembly due to high-grade cement mortars. By carrying out physical experiments using basic, non-professional tools and household products, Ringers found that if our perception of the value of time and labour changes, then so does our relationship with the material. Ringers was left with irregular chunks of bricks with excess mortar, each with its own unique quality. He subsequently explored the Japanese art of Kinsugi ("golden repair", or treating the "breakage and repair as part of the history of an object, rather than something to disguise")[11] as a methodology to piece together the rough edges of these material elements that are typically deemed undesirable. Ringer's project suggests that when we reassemble 'vintage' bricks, there are new opportunities for architectural expression. The problems of re-commodification are exposed, where the idea of 'vintage' becomes a gimmick to add value to the fetishized object.

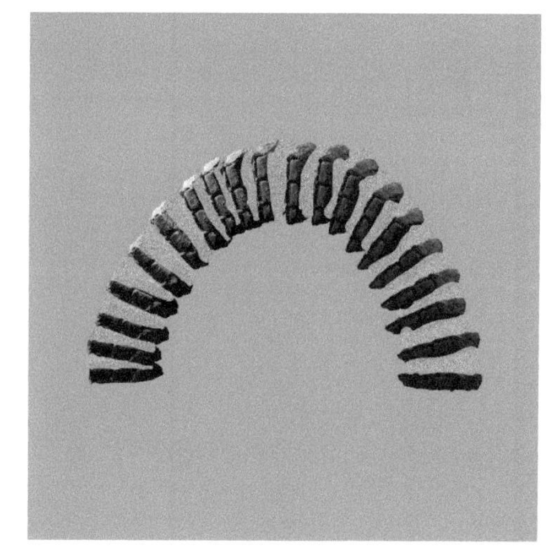

FIGURE 6.5 (left): An imagined 5 kg *Kintsugi Mortar Colour* kit, and (right): a *Kintsugi arch* (day 7). A celebration of the crude patched up aesthetic of manually dismantled brickwork.

Source: Derk Ringers at the School of Re-construction, 2021

Thomas Parker, based in Portugal, started by looking at failing commercial centres, considering that a case might be made for conserving materiality either on the grounds of cultural memory or embodied carbon. Through a careful process of measurement, Parker identifies a palette of re-usable materials with a variety of values relative to their local abundance and the popularity and associations of their visual qualities and finish. Parker concludes with proposals that include a montage of high-quality and low-quality materiality used to form floor surfaces as 'crazy paving'. Here, there is a knowing reappraisal of the suburban domestic driveway mainstay, a realm where DIY aesthetics and personalization reign. By applying these materials in popular mode, Parker's work locates itself in an anti-design aesthetic trend which forces people to reassess a high/low division of commodified material value. Parker makes us wonder that if the economic parameters change, whether this DIY aesthetic could become an industry norm as opposed to a niche eccentricity.

Juxtaposition is a technique that lends itself to *Deep Re-use*. **Julia Flaszynska**, in Austria, worked with 19th-century façade ornaments on a building in Vienna, which had heritage value yet lacked protected status. While watching this building being demolished, Flaszynksa used interviews to analyse the public's emotional experience of the destruction. Flaszynksa then made calculations to quantify similar ornamental fragments that will be erased from Viennese buildings of a similar age and condition.

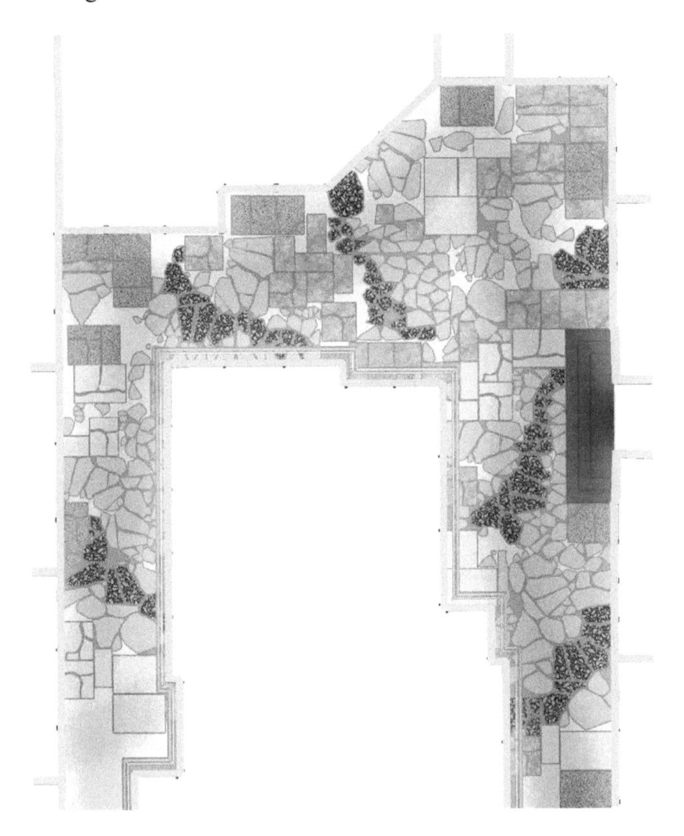

FIGURE 6.6 *Crazy paving* (day 9). Mixed 'value' materials, damaged from the complications of wet fixed dismantling, given new life and a new aesthetic language in a collaged floor surface.

Source: Thomas Parker at the School of Re-construction, 2021

FIGURE 6.7A (left): *Demolition*, Vienna, December 2019

Source: Georg Scherer/wienschauen.at

FIGURE 6.7B (right): *Ornamental climbing wall* (day 8). The lost rubble of Hofmühlgasse 6/Mollardgasse 7 and a graveyard of its demolished fragments are playfully recontextualised in public space as part of a climbing wall.

Source: Julia Flaszynska at the School of Re-construction, 2021. Background image of collage with permission by Sebastian Wahlhuetter/www.wahlhuetter.net

Flaszynksa found that these elements are "an embodiment of longevity and historical meaning". She imagines the emotional impact of relocating some of these ornaments in various public and private environments. One image shows a proportionally massive plaster coil re-used as a secondary structural element (a shelf) – resurrecting its symbolic function but within a domestic environment. In another montage, fragments of ornament are crudely strapped onto the utilitarian metal railings of a bridge like love locks. What gives this project its potency is that the historic or heritage value is heightened by these recontextualizations.

A short history of progress

These student projects draw from a long but often overlooked tradition of *Deep Re-use avant la lettre*. From the columns and friezes of the Arch of Constantine, Rome (315 AD),[12] which re-used imperial reliefs from Trajan, Hadrian and Marcus Aurelius, to more everyday re-use of masonry from dissolved English monasteries, which turned churches into quarries. We can think of these examples as fitting somewhere in the spectrum that spans between ideological re-use and pragmatic re-use. Ideological re-use relates to the objects' cultural value, objects that are re-used because of the meaning of the action, whereas pragmatic re-use refers to an act of resourcefulness or the benefits offered by technical performance. When Cleopatra's Needle was relocated to London,[13] it was a grand statement about power and empire – an ideological re-use.

Deep Re-use is not limited to the realm of grand monumentality and can be found in abundance throughout history as 'common-sense' vernacular self-build. Here, the legacy of material has a strong relationship to the craft of the communities that used them, gaining immaterial

FIGURE 6.8 The inherently highly ideological re-use language of the DIY geodesic domes fabricated from materials such as salvaged car bonnets at *Drop City* (1965–1973).

Source: 'Drop City Panorama' courtesy of the Clark Richert Estate

importance over time. *Deep Re-use* can be unintended, such as in the case of the black steel and mesh railings that line various estates in London. These fences were originally emergency stretchers used by air raid protection officers during the Blitz in 1940–41. During the World War II, many of London's housing estates lost their original iron railings when they were melted down and used for weapon production. With a large stockpile of stretchers following the war, the London City Council had the stretchers welded vertically together and used to replace the missing fencing. It is the juxtaposition, or mis-use of the object, that gives it potency – what was once a pragmatic re-use of material is now a poignant reminder, embedded in London's built fabric.

In latter-20th-century history *Deep Re-use* can be found in experiments of aesthetic subversion and as a counter-culture to industrial material principles and consumerism within the developed world, most notably starting in the 1960s. These associations and legacy still influence our relationship with materiality today, further merging with reactions to the accelerating climate crisis and reassessments of global attitudes towards re-use in reflection of its cultural significance in developing countries, or within non-designed environments.

The good, the bad and the ugly

During the workshop, Parker observed that "conservation on the grounds of heritage, while valid, can only ever represent a fraction of the existing building stock". The relationship between re-use and heritage has both the potential to save or hinder radical transformation of material re-use. Conservation is inherently a sustainable practice, yet it can also be a polarising force, often diverting public attention towards what is 'good' in the built environment and what is not, creating bias and prejudice.[14]

As we move into an era when materials will need to be re-used to a far greater extent because of diminishing raw material and the unviability of mining, we question will 'heritage' be absorbed or, perhaps, exist only to distinguish between preservation and reappropriation? If heritage, as a cultural category in the built environment, becomes irrelevant through absorption, what can we save and re-use from the idea of heritage itself, and how does this also overlap with the increasing number of conversations critiquing the Western architectural canon?

In the UK, we can find new responses to the malleable concept of heritage in the growing trend at the National Trust to try to tell more democratic stories, oral histories, showcasing 'lower status' buildings.[15] Material histories sidestep the traditional bias of text-based histories, which often are written from the privileged perspective.

Underpinning the *Deep Re-use* approach is the idea that culture is upstream from politics sometimes – in the respect that cultural beliefs, whether they are informed by TikTok influencers, authors of novels, lyricists or even architects, can change the wider political discourse. We are on the edge of re-use becoming mainstream. We could assume that this culture would be filtered through the equivalent of TV home improvement shows, where instead of getting over excited about redoing a 3-bed semi-detached in 24 hours, the presenter will marvel at the slow considered placement of a re-used kitchen counter.

The cultural hegemony of the last epoch has been underpinned by capitalist overproduction and dominated by a specific class which demands new and ever larger status symbols – in the

FIGURE 6.9A (left): Capturing the spirit of musical re-use through the labour of sample searching in record stores – Brian Cross's photo used for DJ Shadow's album *Endtroducing*.

Source: B+ for Mochilla.com

FIGURE 6.9B (right): Re-use presenting a "physical manifestation of a thought process"[20] in *Shedboatshed*, Simon Starling, 2005. The installation involving the deconstruction and transformation of a found shed, made into a boat, taken on a journey down the Rhine to a museum in Basel and reassembled back into a shed, bearing the scars from its journey.

Source: Simon Starling, Installation view, *Shedboatshed (Mobile Architecture No. 2)*, 2005, Courtesy of the Artist, The Modern Institute/Toby Webster Ltd, Glasgow

Photo: Kunstmuseum Basel, Martin Bühler

UK, this is typified by virgin marble kitchen islands or wet rooms with abundant glass. Those who now have high cultural capital can change the direction of travel. However, we should be wary of the inevitable co-option of the superficial aesthetic of re-use, which denudes it of its radical potential, as in the case of commodified ripped jeans. Now jeans can be bought pre-worn, given rips by a machine. What was once a symbol of long use or re-use and rejection of the aesthetic of newness is now just another cultural status symbol, unbound to realities of material process. The architectural equivalent of this is aged brick slips being installed at multinational franchises such as Pret a Manger, borrowed from the once counter-cultural warehouse conversion aesthetic – now literally stuck on the surface.

Cultural practitioners, from musicians to visual artists, deploy re-use to create wholly new cultural artifacts. This can be found commonly in video, such as the work of George Barber[16] or Christian Marclay,[17] and in installation work, such as *Shedboatshed* by Simon Starling. The case of music could be considered as having the most popular impact; the electronic group Daft Punk, who have sold over 9 million albums globally,[18] talk about re-use of audio that reflects the freedom that can come with interpretation of the existing: "When you use a sampler, nobody plays on it, so the problem of the ego of the musician is not really there".[19] DJ Shadow's 1996 album 'Endtroducing' pushed audio re-use to an extreme, created almost entirely from samples involving extraction from a vast archive of records. Despite having no obvious 'material' benefit, re-use culture in immaterial expression can influence the perceptions, ideas and actions of wider society towards resourcefulness and innovation in 're-use'.

Material revolutions

Architecture by architects has minimal impact on the material environment in quantitative terms on a global scale, as a small indicator of this, according to the Royal Institute of British Architects, in 2017 only 6% of new homes in the UK were designed by architects. In 2016, 200,000 homes were built in England without the input of an architect[21] – it is within informal and non-professional construction culture where the material revolution will have real impact. Here, the parameters are different, the (self)builders have more time than money, flipping the hierarchies of material choice, and are not driven by the same performance requirements or risks. Improvisational ideas flourish when the maker is the same as the end user, where there is a direct relationship with materiality – a culture of 'ad hocism'.[22] On the date of writing, a YouTube search for videos with the phrase 'DIY' produces over 150 million results. The material impact of *Deep Re-use* within the kitchen or bathroom renovation sector, regularly a DIY staple, gives us an idea of the scale of impact if re-use starts to dominate cultural norms. A 2017 study in America suggested that each year one in ten domestic kitchens are renovated.[23] The material implications are massive.

To have any meaningful impact on carbon consumption, re-use must achieve accessibility at an industrial scale within either professional or non-professional spheres of design. Evolving precedents take various guises internationally, such as the progressive approach of the cooperative RotorDC,[24] concentrating on processing and marketing the redistribution of latter-20th-century building materials in the local context of Brussels. A different example is evolving American deconstruction and redistribution models such as Good Wood,[25] Habitat for Humanity[26] and Details,[27] which operate in the charitable sector with a focus on positive change in material re-use and employment through the social economy. Such third-sector models are not reliant on revenues and can, therefore, afford to process lower value materials, expanding beyond the realm of antiques and challenging the status quo of re-use dominated by concepts of antiquity or established fashions. In the case of RotorDC, this concept "champions a model of materials recovery that resists hippie aesthetics, self-built DIY fads, the fetishization of time-worn surfaces, and the mere circulation of easy-to-reuse modules".[28] With societal change at scale, a plurality of aesthetics can evolve that will cease the need for re-use languages to be claimed as 're-use'.

The opportunity for the overlapping of cultural and technical information within *Deep Re-use* requires new standards that push the cultural challenge further: a model that sits somewhere between the role of a public archive and builders merchant – the museum and the DIY warehouse. A small-scale and imperfect example exists in Porto, Portugal, the 'Banco de Materiais'[29] (Material Bank). This state-run project opened in 2010 with the objective to act as a record of the city's materiality but also as a resource. Aisles of salvaged 'azulejos' (ceramic tiles common on facades in the city) and other decorative or valued construction materials are accessible as if museum exhibits yet also for free supply in the case that a citizen can prove they own a property that is missing elements that are available. However, this model is limited to an agenda of conservation, with a small palette of 'heritage' materials. To engage with the mass market of self-builders and small-scale contractors, we might instead imagine the big-box buildings on the ring roads of cities, currently housing Ikeas, self-storage and Amazon warehouses, being converted into storage of re-usable architectural salvage. It's a very small step – the blurry combination of serious cultural activity mixed with streamlined e-commerce efficiency. The new publicly accessible Science Museum – National Collections Centre, designed by Sam Jacob Studio is an example of curatorial contextualisation and juxtaposition, which have much in common with this imagined re-use depot.[30]

Blending together these last two examples, we can imagine a hybrid 'Re-use Archive'. This archive replaces the go-to staple of public material supply but presents materials organised in a variety of ways that transgress the rules of their original function or value, perhaps as Simone Ferracina imagines when describing the ethos of the 'bricoleur' in the context of architecture and construction, where the organisation of materials "remain in a state of suspension; to exist between value systems and roles".[31] Here, material passports are produced using inquisitive methods, inviting public interpretation and stories, becoming open-source tools. The typical roles of dismantling, handling, cleaning, refurbishment, storage, repackaging, documenting, promotion and resale are complemented by adding the role of cultural interpretation. Instead of a librarian or a sales adviser, we imagine a 'Material Interpreter', who engages the public (amateur or professional) with the materials, providing technical guidance and examples of experimental re-use solutions. They will record more than the dimensions or colour of the virtual brick you can imagine with your eyes closed, nurturing a culture of care for the materials. This role of the Material Interpreter becomes more economical with scale of the operation and the mass of elements, flexible to suit available labour resources at different scales, but ultimately aspiring towards mass job creation. Material value lies in the hands of the participants of a new shared material economy. This form of infrastructure will serve as a resource and memory bank allowing for free interpretation of re-use – recording the minutia and specific qualities that give the objects their cultural value.

This 'Reuse Archive' could take many organisational models. A radical approach, which might have more scalable impact, would be a centralised state model where the materials are mapped and loaned, following the idea of construction within the brackets of a service economy. This environment challenges the association of re-used materials with luxury or marginality; instead, they become a new universal aesthetic language.

FIGURE 6.10 (left): Tile section at the Banco de Materiais, Porto, Portugal, and (right): the Science Museum – National Collections Centre, UK, Sam Jacob Studio – flexible coloured geometric grid awaiting the future organisation of objects in multiple ways. A future 'Reuse Archive' may be a blend of these two spaces: part Aladdin's cave, part museum, part warehouse, part supermarket, part library. A resource for the collective sharing of material history, merging the interpretative freedom created in John Soane's Museum (London) and its mysterious assembly of classical fragments, with the spontaneous and accessible circulation potential of re-use sharing networks such as Freecycle, Craigslist, and OLX.

Source: photograph by Jonny Pugh, SJS/photograph by Timothy Soar

In the professional realm, designers urgently need to think about re-use. This starts with education – the premise of the School of Re-Construction. As Ruth Lang points out, in *Building for Change*, re-use in architecture "usurps dogmatic hierarchies and questions the role of architects within the design process".[32] Thinking critically about how we measure the existing material world is an important foundation. Derk Ringer takes us back to the brick, with a final reflection on the workshop:

> Looking at measurement with curiosity, puts prescribed value judgments into perspective. With this distanced perspective, it is easier to consider neglected and contested values, which might lead to meaningful change in material culture. We started seeing the documentation of material, in my case bricks, as an abstraction that aims to isolate or emphasize a specific quality. The intent behind the abstraction is what became really interesting. Although a specific measurement in itself might be objective, the act of measuring is not. Using measurement as a framework allows us to zoom in on values that challenge common material systems.

Deep Re-use, whether ideological or pragmatic, juxtaposed or harmonised, by an architect or DIY actor, is the key to making re-use part of the mainstream. The emerging pedagogy of climate aware architecture must embrace this cultural challenge.

. . .

With thanks to participants of the workshop: Charissa Leung (Canada), Derk Ringers (the Netherlands), Julia Flaszynska (Austria), Manon Ijaz (UK), Nafisah Musa (Nigeria), Thomas Parker (Portugal), and to visiting external critic Cláudia Escaleira.

Notes

1 https://single-market-economy.ec.europa.eu/industry/sustainability/buildings-and-construction_en, accessed October 1, 2022.
2 *The European Commission Join Research Centre*, News Announcement, Last modified September 9, 2022. https://joint-research-centre.ec.europa.eu/jrc-news/evidence-food-information-empowering-consumers-make-healthy-and-sustainable-choices-2022-09-09_en
3 *"Material Passport"* also known as a *"Product Passport"/"Circularity Passport":* "The majority of current initiatives are limited to defined areas of application . . . To provide a solid data source for a circular built environment, holistic information from different fields is needed". Heinrich, Matthias, and Lang, Werner *Materials Passports – Best Practice*, BAMB (Buildings as Material Banks, 2020), p2.
4 Mounsey, Rosie and Webb, Steve *Climate Action: We Can Put a Block on Brick.* Last modified June 21, 2021. www.ribaj.com/intelligence/structures-sustainability-we-can-put-a-block-on-brick-steve-webb
5 In his book *Ecologies of Inception: Design Potentials on a Warming Planet*, 2022, Simone Ferracina comments that skills in re-use are indispensable, for example, cleaning and laying old bricks, when contrasted with the relatively deskilled labour imposed by the modernist adoption of construction techniques such as reinforced concrete.
6

> Convivial tools are those which give each person who uses them the greatest opportunity to enrich the environment with the fruits of his or her vision. Industrial tools deny this possibility to those who use them, and they allow their designers to determine the meaning and expectations of others.
>
> Illich, Ivan *Tools for Conviviality* (New York: Harper & Row, 1973), extract from chapter II

7 Allwood, Jullian and Cullen, Jonathan *Sustainable Materials Without the Hot Air* (London: Bloomsbury Publishing, 2015), p238.
8 Leopold, David *Karl Marx, Philosophy* (Oxford: Oxford University Press, 2015)
9 Marx, Karl *Capital: A Critique of Political Economy, Vol. 1, Chapter 1, Section 3 the Form of Value or Exchange-Value, Part 4 the Fetishism of Commodities and the Secret Thereof* (Moscow: Progress Press, 1867).
10 "Today, prospecting methods are based on advanced techniques and scientific methods", yet also rely on a "great talent for observation", as described by Ghyoot, Michaël, Devlieger, Lionel, Billiet, Lionel and Warnier, André *Déconstruction et réemploi. Histoires, tendances et perspectives* (Lausanne: EPFL Press, 2018).
11 Kintsugi, In *Wikipedia.* Last modified October 28, 2022. https://en.wikipedia.org/wiki/Kintsugi
12 "the entirety of its roughly 16,000 marble blocks were derived from earlier monuments": Barker, Simon speaking at the Architectural Association event *'(Re)Building with Stone: Ashlars, Spolias, Quarries and Cities'* (London: Architectural Association, January 28, 2022).
13 Cleopatra's Needle is an ancient Egyptian obelisk now located on the Victoria Embankment, London. Inscribed by Thutmose III and later Ramesses II, the obelisk was moved to Alexandria in 12 BC, where it remained for nearly two millennia before it was presented to the United Kingdom in 1819 by the ruler of Egypt and Sudan Muhammad Ali.
14 Illustrating this point, the critic Rafael Gómez-Moriana discusses the Flores & Prats project Sala Beckett in his article 'Circle of Life' (*The Architectural Review*, December 2021) as a methodology that rejects 'old' versus 'new' binary ideology in favour of a language of 'renewed old':

> The Venice Charter established the heritage restoration guidelines still in use today. Article 12 explains why, in many historic building restorations, a sharp contrast is emphasised between old and new: a visual trope that has come to express a certain idea of historical rupture or discontinuity between past and present. Never mind that most historic buildings are, themselves, products of multiple interventions and adaptations throughout the ages. For some reason, historical changes are never considered 'falsifications', whereas contemporary ones are – which is why the latter must always be 'distinguishable'.

15 For example, Rainham Hall opened to the public in 2015 with stories about all its historic inhabitants rather than a single famous character www.studioweave.com/projects/rainham-hall/

16 The rapid editing of recycled video material in *The Greatest Hits of Scratch Video, Volume One & Volume Two* (Barber: George and Various Artists, 1984–1985).
17 For example *The Clock*, 2010, a 24-hour long installation made from a montage of thousands of film and television images of clocks, edited together so they show the actual time.
18 https://bestsellingalbums.org/artist/2807, accessed October 30, 2022.
19 'Daft Punk', Reesman, Bryan, Mixmag, 1st October 2001, as referenced in *Domestic Discovery or Sampling and Synthesising With Daft Punk*, Geoff Shearcroft, *P.E.A.R Paper for Emerging Architectural Research*, Issue 3, 2011.
20 www.tate.org.uk/whats-on/tate-britain/turner-prize-2005/turner-prize-2005-artists-simon-starling
21 www.architectsjournal.co.uk/news/most-new-housing-so-poorly-designed-it-should-not-have-been-built-says-bartlett-report, accessed October 30, 2022.
22 Ghyoot, Michaël, Devlieger, Lionel, Billiet, Lionel and Warnier, André. *Déconstruction et réemploi. Histoires, tendances et perspectives* (*Pu Polytechniqu*, 2018) (making reference to the term adhocism in: Jencks, Charles and Silver, Nathan *Adhocism. The Case for Improvisation* (Cambridge, MA: MIT Press, 1972).
23 In the case of kitchens, almost half of these were budgeted at $15,000 or more www.prnewswire.com/news-releases/americans-love-their-kitchens-and-baths-nkba-research-pegs-industry-value-at-134-billion-300387346.html
24 https://rotordc.com/, accessed September 23, 2022.
25 https://urbanwoodgoods.com/, accessed October 30, 2022.
26 http://www.habitat.org/restores, accessed November 2, 2022.
27 https://details.org/, accessed October 30, 2022.
28 Ferracina, Simone *Ecologies of Inception: Design Potentials on a Warming Planet*, Chapter 'Architectural Bricolage' (London: Routledge, 2022).
29 https://museudacidadeporto.pt/estacao/banco-de-materiais/, accessed October 30, 2022.
30 www.samjacob.com/portfolio/national-collections-centre/, accessed October 30, 2022.
31 Ibid.
32 Lang, Ruth *Building for Change – The Architecture of Creative Reuse* (London: Gestalten, 2022).

7

USELESS 2 'BANQUETING IN USELESS BUILDINGS'

Andre Viljoen

What follows this text is a facsimile copy of the *Recipe Book for Banqueting in Useless Buildings*, produced during the School of Re-Construction by participating students.

The pedagogic approach for this two-week long programme was developed in collaboration with Inês Neto dos Santos, drawing on my architectural design research and Inês' arts practice. My research explores the ways food systems, including urban agriculture, can be reintegrated into cities; Inês is a multidisciplinary artist utilising performance, installation, and social sculpture to investigate food in its intertwined socio-political, cultural, and ecological dimensions. Students were provided with a framework that used recipes and cooking as methods for systematically evaluating the re-use potential within redundant buildings and for a more open-ended exploration of what that re-use could be.

The scenario given to students was of its time, referencing the lived experience of the COVID pandemic:

- the demand for office buildings is declining due to the attractiveness of home working
- the well-being benefits of access to nature, unprocessed food, and physical meeting are being recognised as a by-product of the "lockdown" experience

Because this work was done virtually and participants were in different cities, each student was asked to identify an office building in their location to which they could gain access. Working as an interdisciplinary team, proposals were made for inserting an urban farm and banqueting space into each office building, after which one building was selected for further development.

We had four redundant buildings to work with, in Rabat, Morocco; Bialystok, Poland; Kaunas, Lithuania; and Vienna, Austria. These were imagined to be in one virtual city, and students proceeded to categorise them as material banks and, at the same time, articulate their qualities and potentials. It was at this point that Inês ran a session titled "Recipe for a building – repurposing spaces through repurposing recipes; food and gestures, context and narrative", looking critically at the construction of recipes (and buildings) while exploring written, spoken, gestural, and visual language surrounding food. It touched upon performance and narrative and discussed

DOI: 10.4324/9781032665559-12
This chapter has been made available under a CC-BY-NC-ND 4.0 license.

how using stylistic devices could help tell a story and create a framework, whether linear or nonlinear. Initially students wrote a recipe for the traditional use of their chosen building:

> Think of it objectively and write a recipe for its traditional use. Be specific and include as many details as possible. Write in a direct, objective manner, avoid writing emotionally about it. Follow a traditional recipe format – A title, list of "ingredients", a set of steps/instructions to follow and a description of the final result.

After further discussion and exercises students were asked to break away from instructional writing and bring together the two 'spaces' of food plus built environment starting with the expression *"And then I saw . . ."*, to describe a scenario merging food and the building of their choice:

> Imagine a banquet, featuring the dish in your chosen recipe – consider smells, textures, colours, the people you are with, the other species that might surround you – describe them too. Avoid instructions and write emotionally, sensorially about this scenario. When does it happen? Today, tomorrow, 10 years from now, 50 years from now? Who eats at this banquet, what do the ingredients look/smell/taste like and where do they come from? Who cooks?

From this point onwards, we merged architectural design with cooking and conversation, imagining a repurposed building that accommodate spaces for agriculture, cooking, and eating.

The student's *Recipe Book for Banqueting in Useless Buildings* starts by describing the U2 building in Vienna, which was selected to develop in detail. All alterations and additions use components from the material banks quantified earlier in the process. After the introduction of the U2 building, the recipe book documents the design process, recipe writing, and banqueting.

We enjoyed the process and hope you enjoy reimagining it.

Participating students Ugne Neveckaite, Soukaïna Lahlou, Natalia Hryszko, and Natasha Hromanchuk.

For further information about the architectural and arts practices underpinning this programme see:

Continuous Productive Urban Landscapes (CPULs) Designing Urban Agriculture For Sustainable Cities. A. Viljoen editor and with K. Bohn principal author, Published Architectural Press, ISBN 0750655437: RIBA book of the week, February 2005, and cited as suggested reading by the Royal Commission on Environmental Pollution, for their investigation into Urban Development (2005).

Second Nature Urban Agriculture: Designing Productive Cities. Editors A. Viljoen and K. Bohn. Published by Routledge. Content by Andre Viljoen and Katrin Bohn, with contributions by invited authors. This book is a sequel and companion volume to the 2005 book, *Continuous Productive Urban Landscapes (CPULs) Designing Urban Agriculture For Sustainable Cities*. It was awarded the RIBA President's Award for outstanding university located research.

ines-ns.com. (n.d.). *Home – Inês Neto dos Santos*. [online] Available at: https://ines-ns.com/Home [Accessed 2 Aug. 2023].

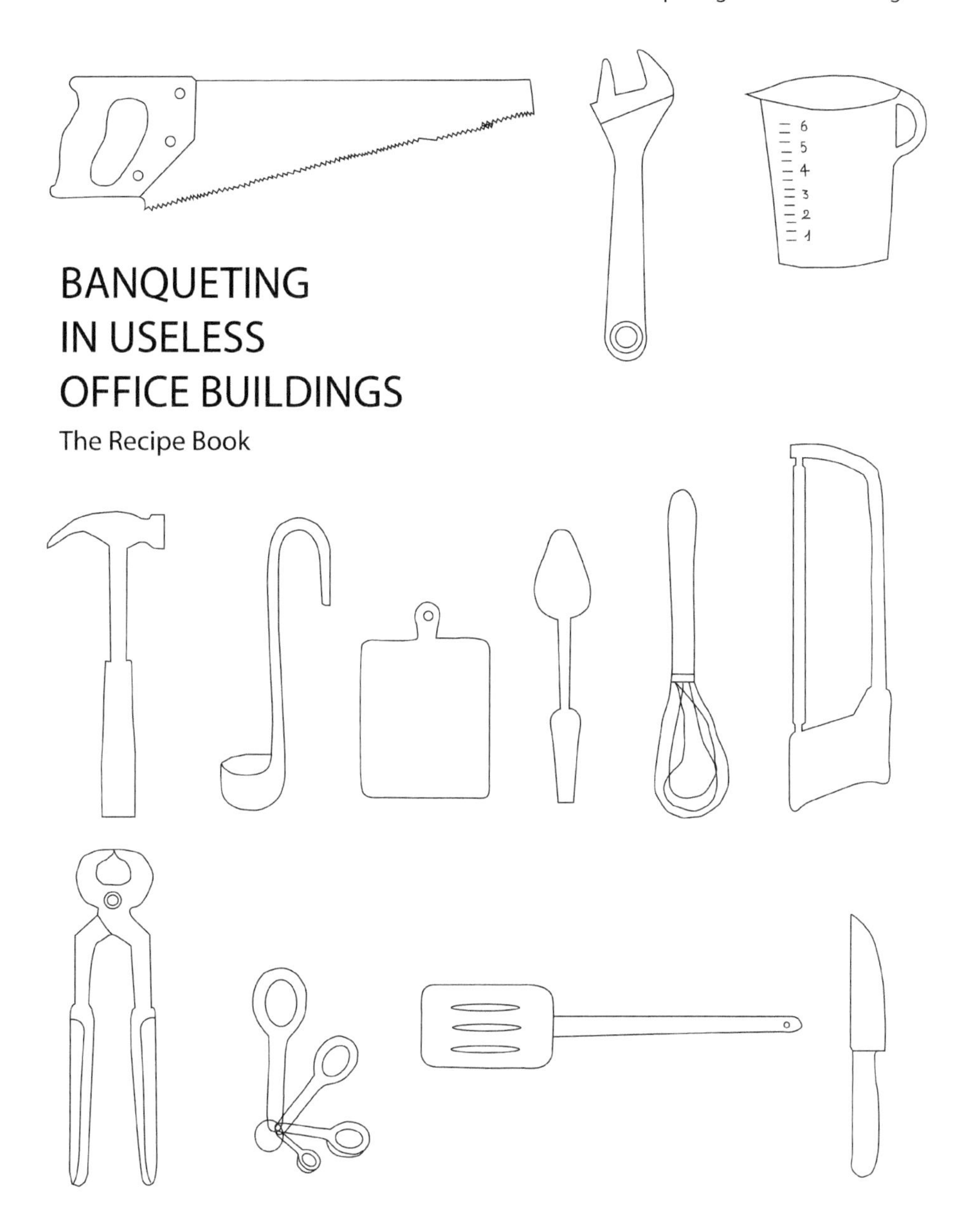

BANQUETING IN USELESS OFFICE BUILDINGS

The Recipe Book

By Natasha Hromanchuk, Natalia Hryszko, Soukaïna Lahlou and Ugne Neveckaite
Prof. Andre Viljoen & Inês Neto Dos Santos
The School of Re-Construction, University of Brighton

U4 building; Vienna, Austria

Table of contents:

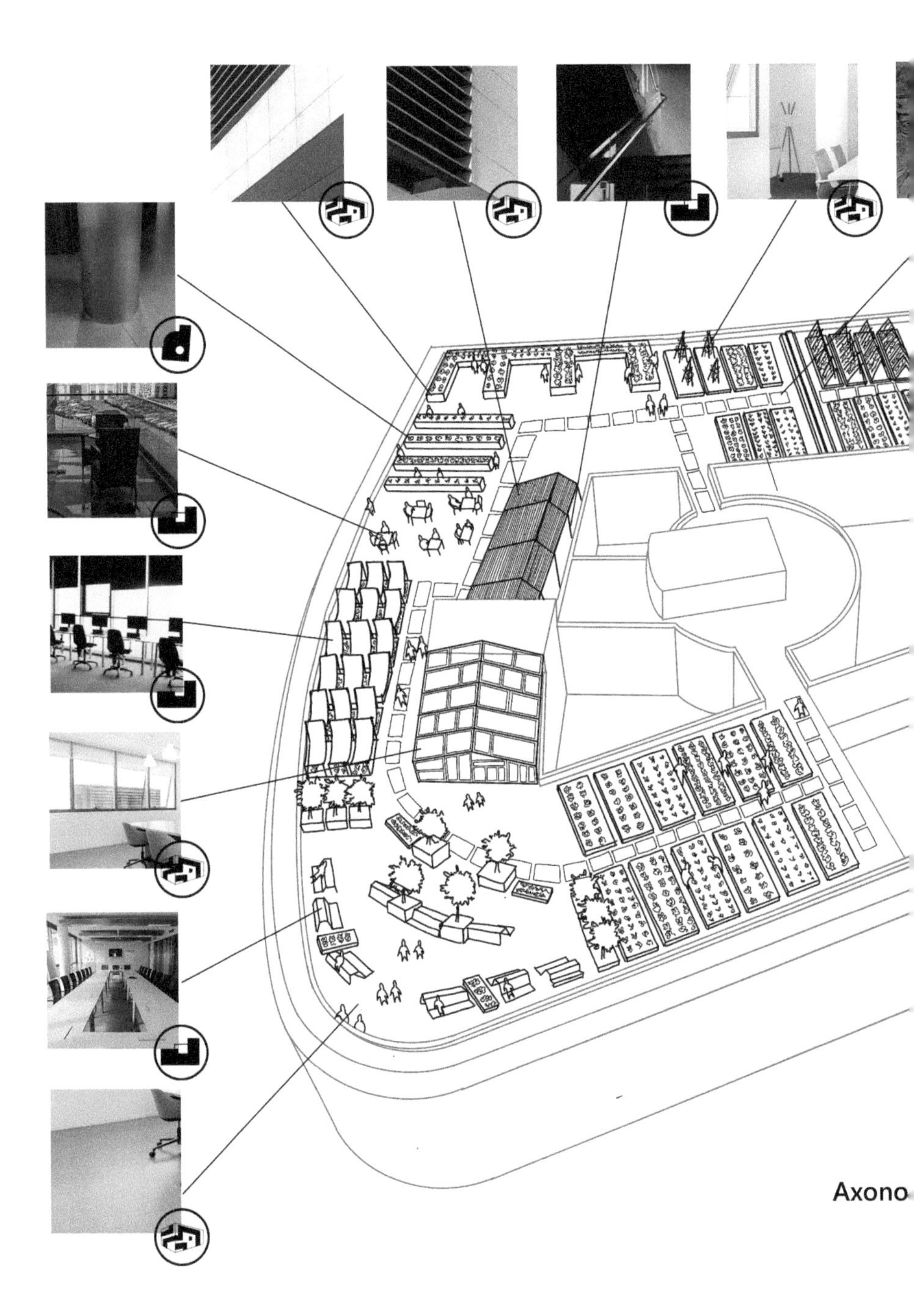

Axono

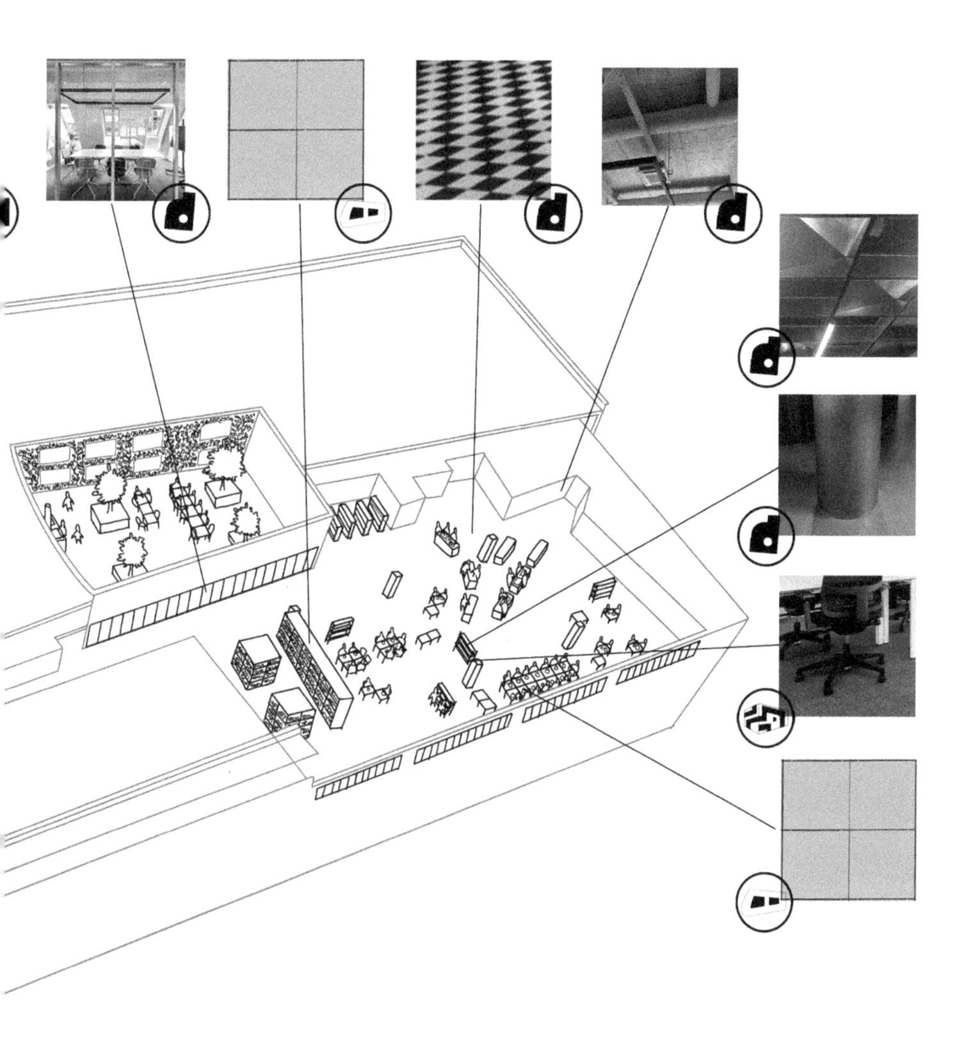

view of the proposed urban farming and banqueting spaces in U4 building

Vienna, Austria

Kitchen

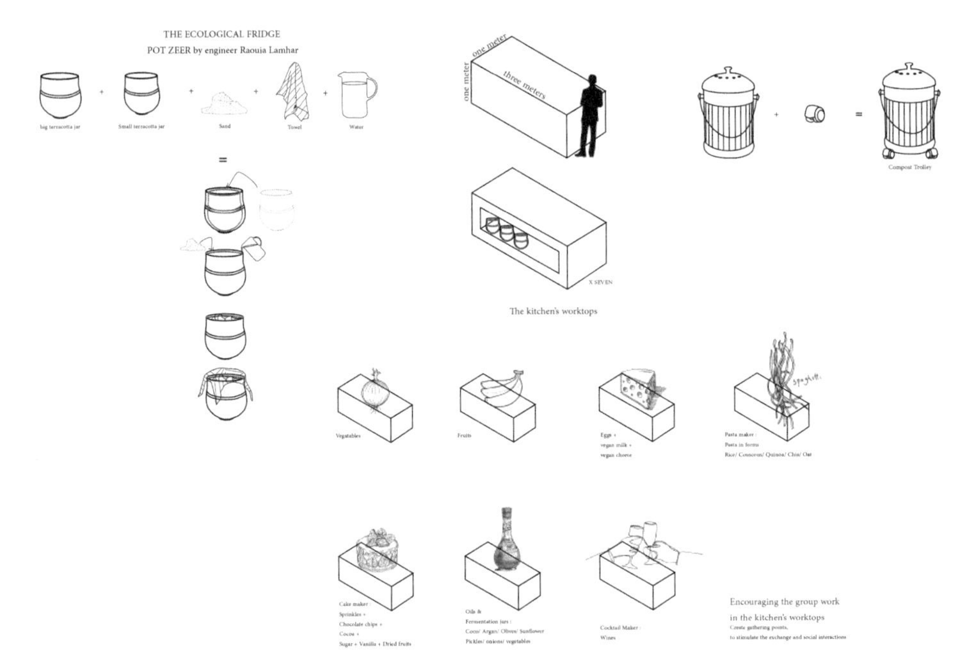
THE ECOLOGICAL FRIDGE
POT ZEER by engineer Raouia Lamhar
big terracotta jar
Small terracotta jar
Sand
Towel
Water
one meter
one meter
three meters
X SEVEN
The kitchen's worktops
Compost Trolley
Vegatables
Fruits
Eggs +
vegan milk +
vegan cheese
Pasta maker :
Pasta in forms
Rice/ Couscous/ Quinoa/ Chia/ Oat
Cake maker :
Sprinkles +
Chocolate chips +
Cocoa +
Sugar + Vanilla + Dried fruits
Oils &
Fermentation jars :
Coco/ Argan/ Olives/ Sunflower
Pickles/ onions/ vegetables
Cocktail Maker :
Wines
Encouraging the group work
in the kitchen's worktops
Create gathering points,
to stimulate the exchange and social interactions

Banqueting space

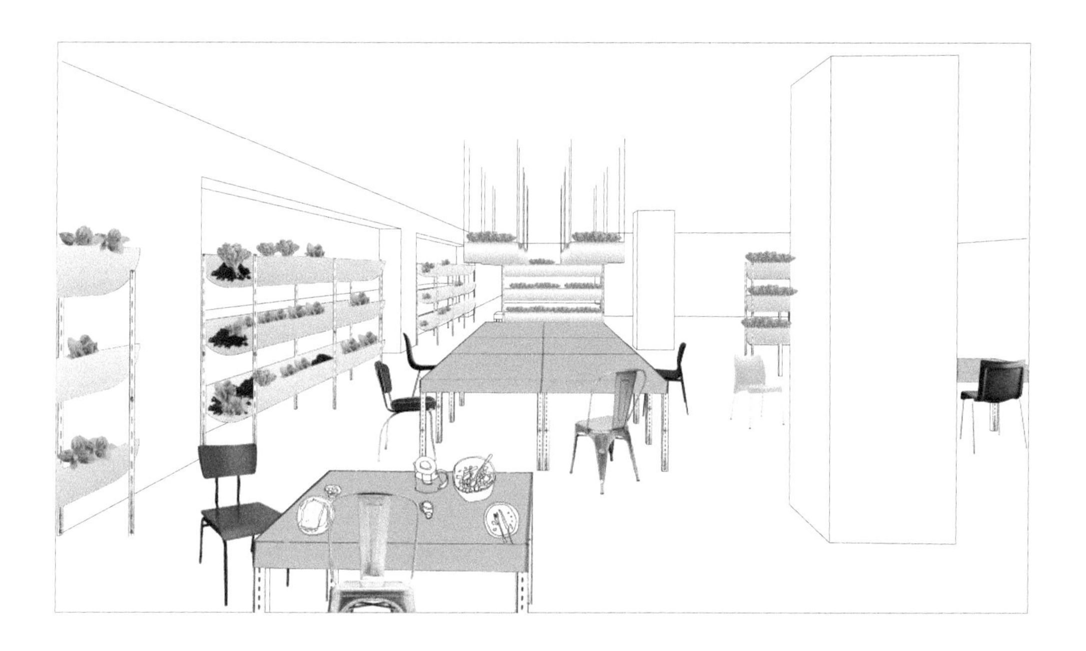

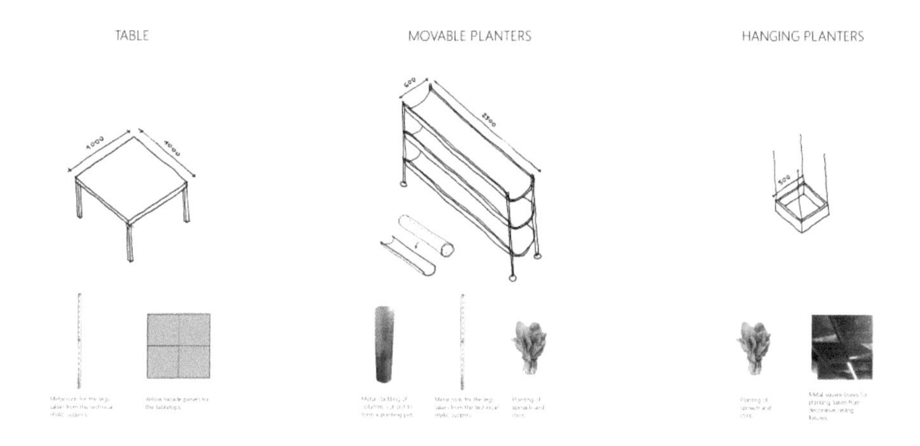

Rooftop greenhouse and urban farming

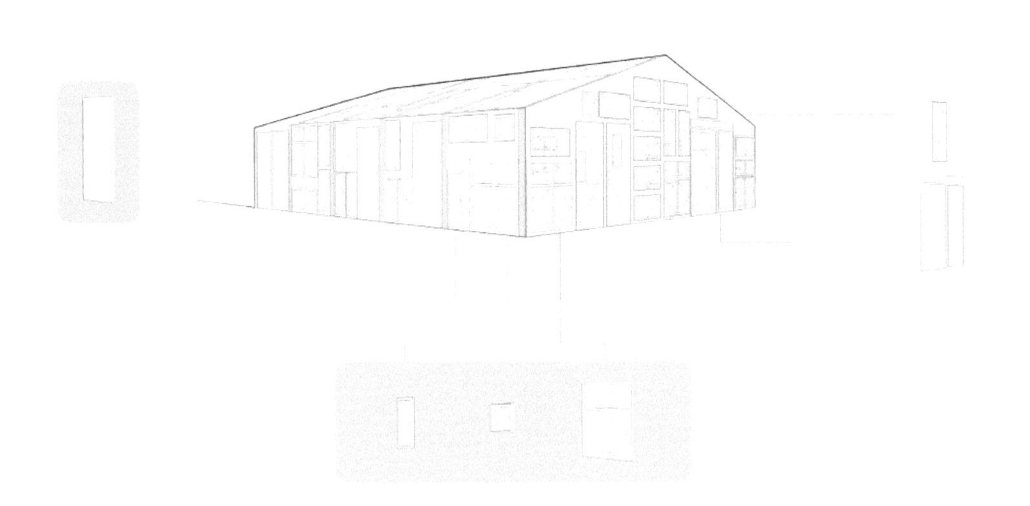

Rooftop seating and tea drinking area

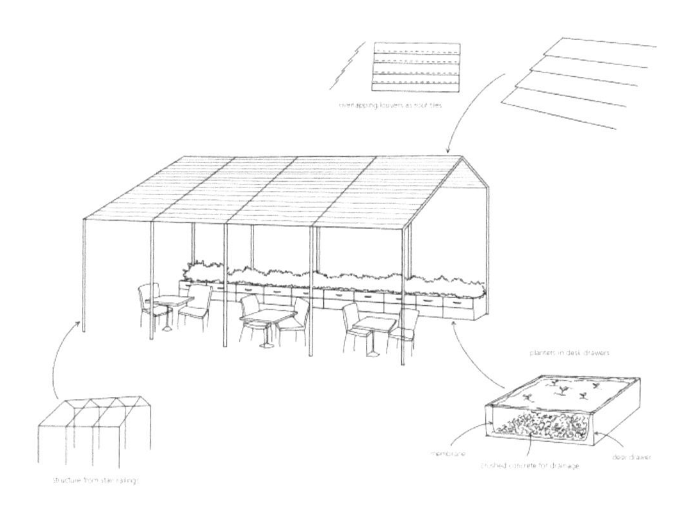

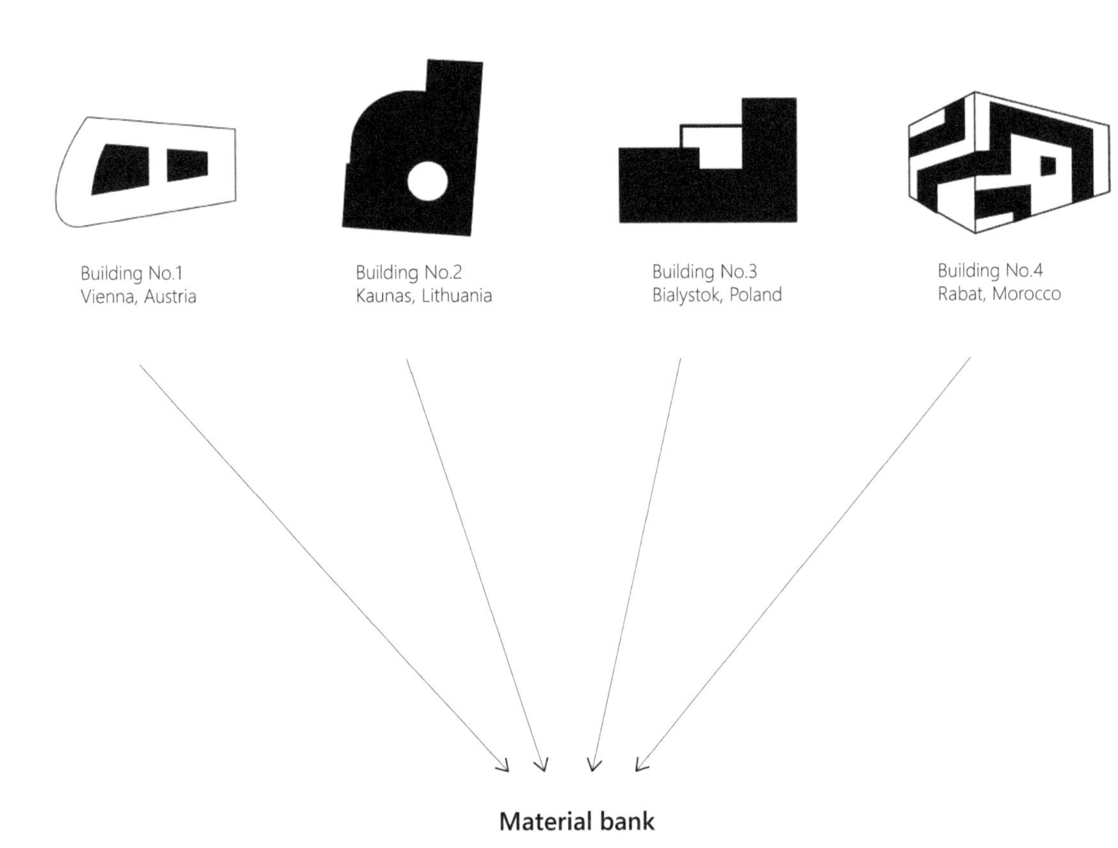
Process mapping
Building No.1
Vienna, Austria
Building No.2
Kaunas, Lithuania
Building No.3
Bialystok, Poland
Building No.4
Rabat, Morocco
Material bank

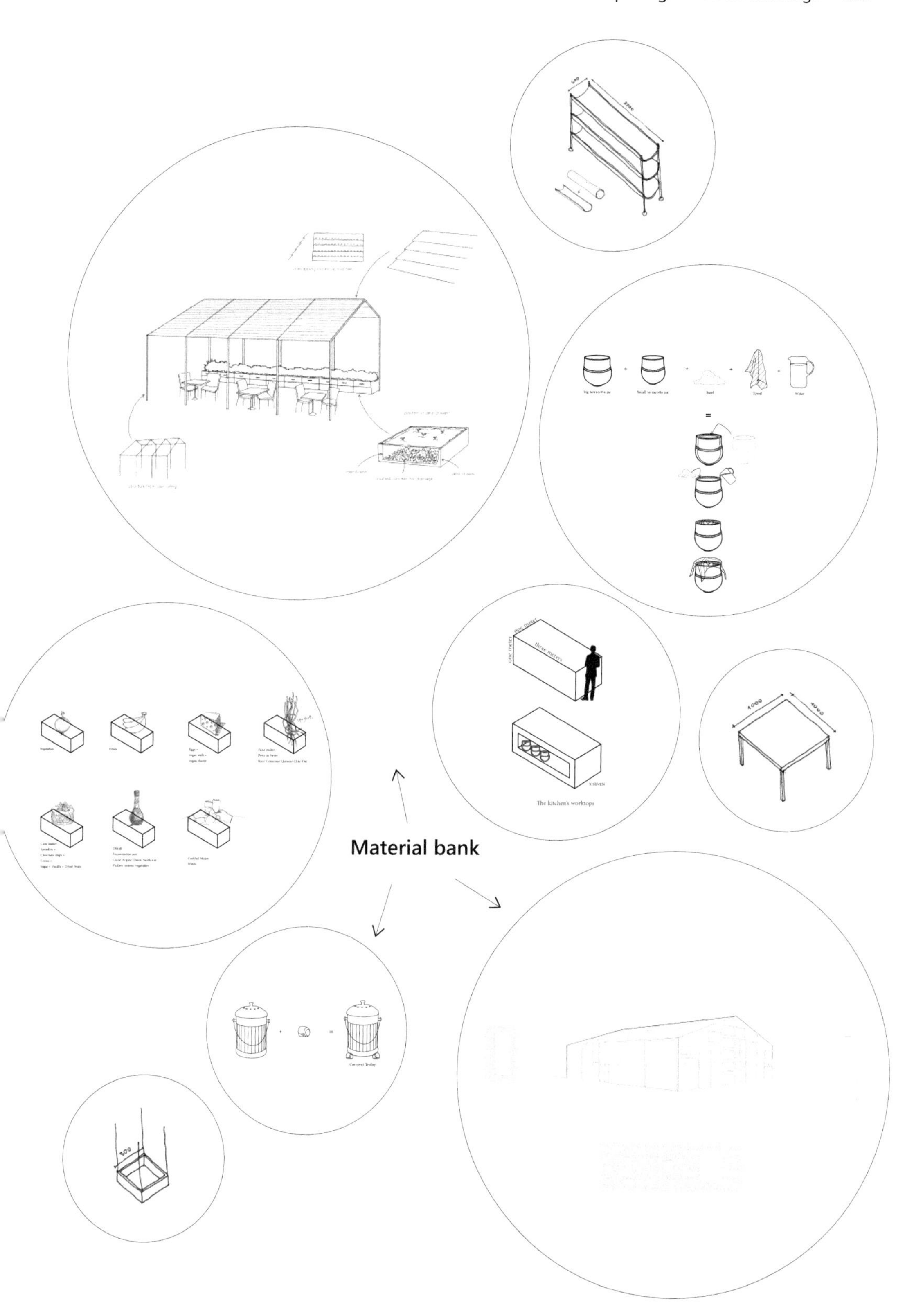
Small terracotta jar
Sand
Towel
Water
one meter
one meter
three meters
X SEVEN
The kitchen's worktops
1000
4000
500
Compost Trolley
Material bank

Material bank

Name: "U4 Center".
Location: Vienna, Austria.
Date: 1980 and 2002
Previous / existing use: Parking, bus station, metro station, retail, fitness, offices, bars, disco.

CATALOG - EXTERIOR

STRUCTURAL
- structural profile
- structural column

DECORATIVE

PANEL
- corrugated metal panles
- white panels
- white panels
- black panels
- corrugated metal panels horisontal
- yellow panels

OTHER
- letters
- blue awnings
- additional roof
- sign
- metal blinds

TECHNICAL
- ventilation panels
- exterior light
- ventilation

DOOR AND WINDOWS

DOORS
- doors entrence tech
- doors entrence
- doors entrence tech

WINDOWS
- window with the cover
- openable windows
- window
- rounded window
- window panels
- window panels
- window

STAIR

CATALOG - INTERIOR

STRUCTURAL
- structural comuns
- structural walls

DECORATIVE
- blue horisontal
- floor tiles
- floor tiles shiny
- suspended ceiling panels

FURNITURE

TECHNICAL
- ventilation
- ceiling lamp
- lift
- light panles
- heater

DOOR AND WINDOWS
- doors office entrence
- white office doors
- doors with a rounded

Tante Fanny
FRISCHER
Blätterteig
Zwiebeln
Das steht auf dem Ei

Name: "Workland" co-working space.
Location: Kaunas, Lithuania.
Date: 1900
Previous / existing use: Bakery, cinema, photo atelier, night club, bank, apartments.

CATALOG - INTERIOR

DECORATIVE & CLADDING			FURNITURE	TECHNICAL	DOOR AND WINDOWS	STAIRS & RAILINGS
FLOOR	CEILING	WALL/PARTITION/ETC				
Carpet flooring	Metal ceiling on the ground floor foyer	Glass partitions	Soft velvet sofas	Technical tubes	Timber doors	Metal railings
Flooring panels		Timber panels,	Timber storage unit	Technical appliances on the roof	Atrium windows	Plastic cladding
Black and white tiles		Metal cladding for all the strucutral columns		Metal glazed elevator	Attic windows	
		Curtain-soft textural material				
		Wall tiles in the toilets				

Name: "Swietojanska Offices".
Location: Bialystok, Poland.
Date: 2016.
Previous / existing use: Offices.
Site surface: 1400 m2.
Architect: W&W Architekci.

CATALOG - EXTERIOR AND INTERIOR

DECORATIVE	FURNITURE		TECHNICAL	DOOR AND WINDOWS	RAILINGS
INORGANIC	ORGANIC	INORGANIC			
Perforated ceiling panels, brown	Bar table, timber	Couch leather white	Cake fridge	Clear glass restaurant door	Metal stair railings
Perforated ceiling panels	Restaurant tables	Table reception	Ventilation	Frosted glass single doors	Metal and glass railings
Suspended ceiling panels white	Shelves restaurant, timber, (2)	Couch leather brown	Ventilation	Clear glass double doors	Metal stair railings
Frosted glass panels	Timber and metal legs desks	Armchair leather brown	ATM	Palm tree glass double doors	Metal and glass stair railing
Marble floor tiles	Office chairs	Bar stools, black, (3)	Lifts	Palm tree glass double doors (2)	Metal and glass railing
Curved marble tiles on columns	White tables	Mirror gold frame	TV (many)	Toilet cubicle plastic doors	
Grey striped marble tiles	Conference room tables	Glass sheet information board	Metal tubes	non-openable restaurant window	
White marble tile		Marble tile reception desk	Water dispenser	Palm tree non-openable tall windows	
Frosted glass panels on stairs		Trash cans	Fridge	curtain wall non-openable windows	
Carpet navy blue		White MDF kitchen cabinets	Hand dryers	curtain wall openable windows	
Marble backsplash tiles		Artwork in glass frames (3)	Sinks and taps		
Marble bathroom		Office chairs leather	stair lift		
Glass sheets office dividers		Conference room table			
Sound absorbing ceiling (?)		Office chairs			
Light grey marble floor tiles		Metal blinds			
Blackstone tiles		Mirror sheets big (bathrooms)			
Terracota tiles white		Marble countertop			
Terracota tiles grey		Restaurant chairs pleather			
Marble tiles		Artwork black frames, square			

BUTTON MUSHROOMS
12 Aug
200g
BABY
READY TO EAT
by Sainsbury's
13 Aug
DELICATE & MILD
GROUND CUMIN
O
OREGANO
P
SMOKED PAPRIKA
TESCO
PLAIN FLOUR

Name: "Ryad center".
Location: Rabat, Morocco.
Date: no exact date - recent.
Previous / existing use: Offices for public or private companies, offices for rental.
Site surface: 8319 m2.

CATALOG - EXTERIOR

DECORATIVE & CLADDING		FURNITURE	TECHNICAL	DOOR AND WINDOWS
ORGANIC	INORGANIC			
wooden mural frames	Polished concrete floor	Wooden tables for group reunions (1,2 x 2,5)	Ceiling lights (squares)	Fabric sun shade
wooden cladding	Plasterboard ceiling	wooden shelves		Aluminium window framing
Carpets (a lot)	Tinted glass panels	stuffed chairs (sofa style)		Glass
Plants		High bar chairs		wood
wooden cupboard		wooden table		
		sofa chairs		
		bar table		

CATALOG - INTERIOR

STRUCTURAL	DECORATIVE	TECHNICAL	DOOR AND WINDOWS
concrete	Glass curtain		
stone	shutter		
	stone blocs		

COUSCOUS
DARI

Recipe for the office builing

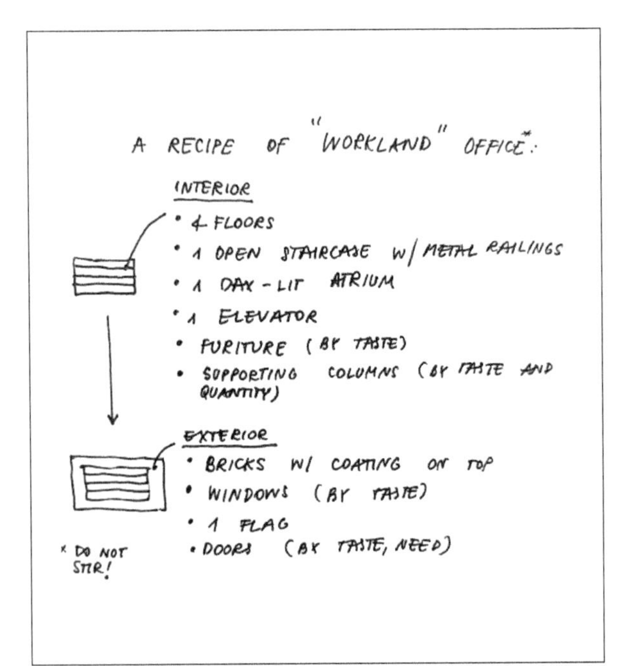

A RECIPE OF "WORKLAND" OFFICE*:
INTERIOR
• 4 FLOORS
• 1 OPEN STAIRCASE W/ METAL RAILINGS
• 1 DAY-LIT ATRIUM
• 1 ELEVATOR
• FURITURE (BY TASTE)
• SUPPORTING COLUMNS (BY TASTE AND QUANTITY)
EXTERIOR
• BRICKS W/ COATING ON TOP
• WINDOWS (BY TASTE)
• 1 FLAG
• DOORS (BY TASTE, NEED)
* DO NOT STIR!

RYAD CENTER - RABAT 10100

For 1 independent complete building :
• A lot of Stone
• A lot of glass
• A lot of concrete
• A little bit of wood

When walking in the middle of Ryad district, don't get lost looking at the big companies buildings. Keep your hat on and continue in a right direction towards the "place carré".
Once in the middle of this place, if you see 8 look-alike buildings you are in the right spot.
Enjoy your balade in between baby palm trees and plants. If you are tired have a sit on the multiple benches and if you are feeling hot, freshen up in the fountain.

U4 CENTRE RECIPE

you need:
• your location
• the aim

You can approach the building from the sides. You will see it immideatly because of the size and colors. Walk along it or go directly from the side streets, it depends on your location.
Indicate colorful doors and go inside. You can decide either to go down with the escalator (to the metro), or to go up with lift (to the offices). Also you can just stay there to shop or to take a bus

SWIETOJANSKA OFFICES

Ingredients:

• Terracotta tiles in large quantities
• Marble tiles in large quantities
• Double doors
• Big white leather couch
• Coffee table
• Reception desk clad in marble tiles
• Three big lifts

Preparation:

Approach the building from Swietojasnka street. Do no enter from the parking side if possible. Go up the stairs clad with marble tile leading up to the front door. Oper the double glass door and walk past the ATM.
When you want to go to the office, walk through the big opening in the wall clad in glass panels and go towards the lifts.
When the lift arrives on the desired floor, exit and walk on the carpeted floor towards the big open space.
Enjoy your day in the office!

Banqueting in the office building- *creative writing exercise*

“ *And then I saw - the crusty dough, sprinkled with big shiny pieces of sugar, brown and with little spillovers of cherry juice on the sides. It smelled fresh and once bitten, the heat of the pastry did warm up the mouth with the scent of cherries, which were tender and big. Once you start eating on those, it is hard to stop and the cherry juice starts dripping down your sleeve like an undercover murderer.* ”

“ *"Hurry up they've started".*
"Started what ?" I've thought to myself. I followed his hasty steps around the fountain and when I looked up I saw something. Something I've never seen before. Something I've felt before.
It was happening in between the laughter, the utensils scratching the cooking pots, the drops of water in the fountain, the human voices, the birds singing, the peace and the chaos. Someone in my back patted my shoulder. I looked back and I see two shinning eyes, there was a young girl inviting me to taste her new bread. ”

“ *It is about the time for nature to reclaim the building back. The seeds of tomatoes are falling out while cutting, tomatoes are overripe and soaking, spilling insides on the cutting board. Community collects this seeds and dries them. Next year they will be planted. The day is coming to an end and the sharp taste of onions stays in your mouth the longest. How do you experience surroundings while eating? The muted noise of cars in the distance goes slowly down and wind spreads flavours around us* ”

“ *Behind the delicate, deliciously smelling cloud of smoke, I could see two chiefs working around a gas stove placed where there used to be a computer and a cutting board covered with a colourful mixture of tiny pieces of vegetables. Rhythmical noise made by the knife hitting the cutting board and sending to the side neatly cut pieces of bell peppers, soft grey mushrooms and evenly cut onion, was resonating from the smooth marble tiles omnipresent in the reception. The second chef standing in front of the stove was slowly stirring the vegetables in a pan. The deliciously smelling steam was forming spirals under the ceiling and merging in my eyes with the white and grey veins on the marble covering the columns and walls.* ”

Virtual banquete

Starter: couscous salad

Main course: home-made pasta with vegetables

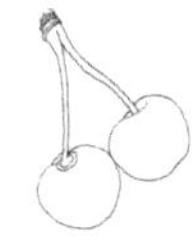

Dessert: cherry pastries

Drink: Moroccan mint tea

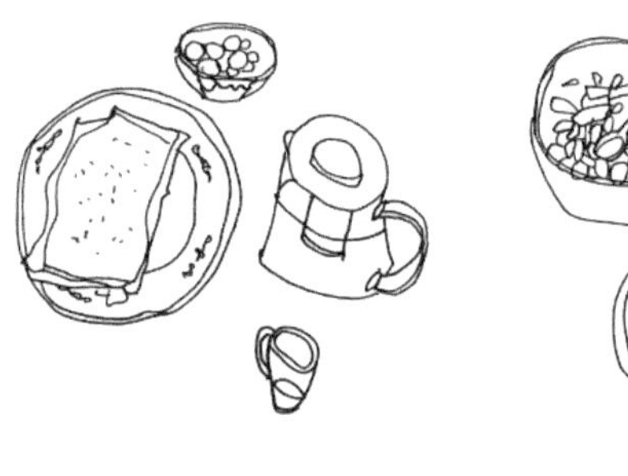

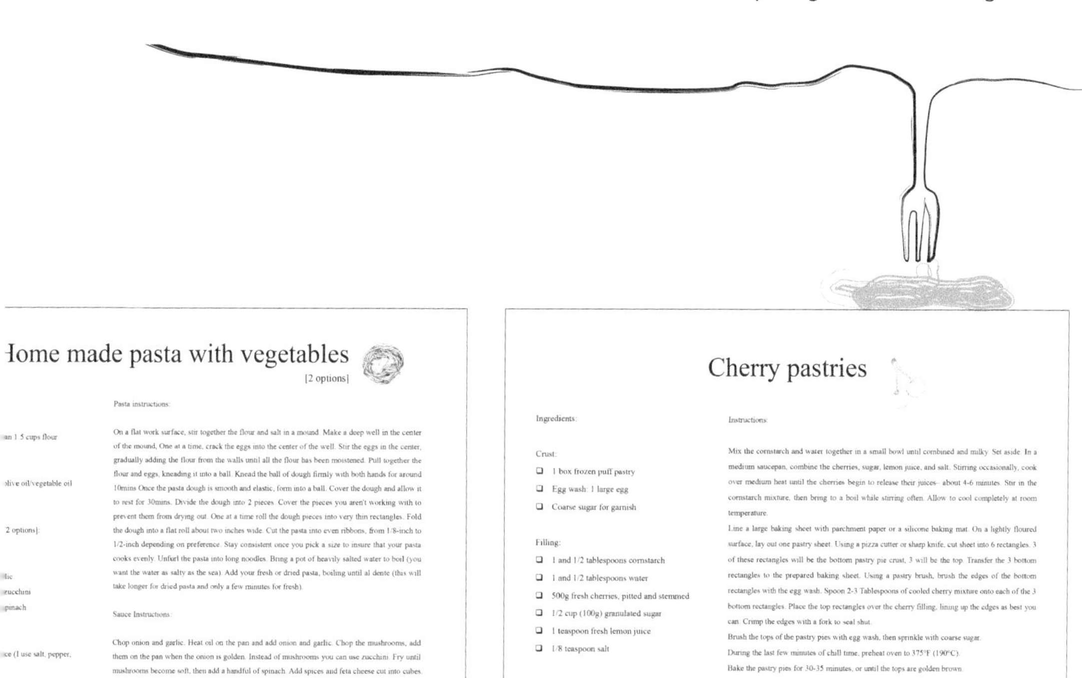

Iome made pasta with vegetables

[2 options]

an 1 5 cups flour

olive oil/vegetable oil

2 options]:

lic

zucchini

pinach

ce (I use salt, pepper,

Pasta instructions:

On a flat work surface, stir together the flour and salt in a mound. Make a deep well in the center of the mound, One at a time, crack the eggs into the center of the well. Stir the eggs in the center, gradually adding the flour from the walls until all the flour has been moistened. Pull together the flour and eggs, kneading it into a ball. Knead the ball of dough firmly with both hands for around 10mins Once the pasta dough is smooth and elastic, form into a ball. Cover the dough and allow it to rest for 30mins. Divide the dough into 2 pieces. Cover the pieces you aren't working with to prevent them from drying out. One at a time roll the dough pieces into very thin rectangles. Fold the dough into a flat roll about two inches wide. Cut the pasta into even ribbons, from 1/8-inch to 1/2-inch depending on preference. Stay consistent once you pick a size to insure that your pasta cooks evenly. Unfurl the pasta into long noodles. Bring a pot of heavily salted water to boil (you want the water as salty as the sea). Add your fresh or dried pasta, boiling until al dente (this will take longer for dried pasta and only a few minutes for fresh).

Sauce Instructions:

Chop onion and garlic. Heat oil on the pan and add onion and garlic. Chop the mushrooms, add them on the pan when the onion is golden. Instead of mushrooms you can use zucchini. Fry until mushrooms become soft, then add a handful of spinach. Add spices and feta cheese cut into cubes.

Cherry pastries

Ingredients:

Crust:

- 1 box frozen puff pastry
- Egg wash: 1 large egg
- Coarse sugar for garnish

Filling:

- 1 and 1/2 tablespoons cornstarch
- 1 and 1/2 tablespoons water
- 500g fresh cherries, pitted and stemmed
- 1/2 cup (100g) granulated sugar
- 1 teaspoon fresh lemon juice
- 1/8 teaspoon salt

Instructions:

Mix the cornstarch and water together in a small bowl until combined and milky. Set aside. In a medium saucepan, combine the cherries, sugar, lemon juice, and salt. Stirring occasionally, cook over medium heat until the cherries begin to release their juices– about 4-6 minutes. Stir in the cornstarch mixture, then bring to a boil while stirring often. Allow to cool completely at room temperature.

Line a large baking sheet with parchment paper or a silicone baking mat. On a lightly floured surface, lay out one pastry sheet. Using a pizza cutter or sharp knife, cut sheet into 6 rectangles. 3 of these rectangles will be the bottom pastry pie crust, 3 will be the top. Transfer the 3 bottom rectangles to the prepared baking sheet. Using a pastry brush, brush the edges of the bottom rectangles with the egg wash. Spoon 2-3 Tablespoons of cooled cherry mixture onto each of the 3 bottom rectangles. Place the top rectangles over the cherry filling, lining up the edges as best you can. Crimp the edges with a fork to seal shut.

Brush the tops of the pastry pies with egg wash, then sprinkle with coarse sugar.

During the last few minutes of chill time, preheat oven to 375°F (190°C).

Bake the pastry pies for 30-35 minutes, or until the tops are golden brown.

banquet was documented with more visual details in a video we did for the workshop but also through zoom call dings as the cooking and the eating part were done in group through our screens.

Réunion

Enregistré par : Andre Viljoen

Réunion

Enregistré par : Andre Viljoen

Cooking at 1 pm (UK Time)

Eating at 7 pm (UK Time)

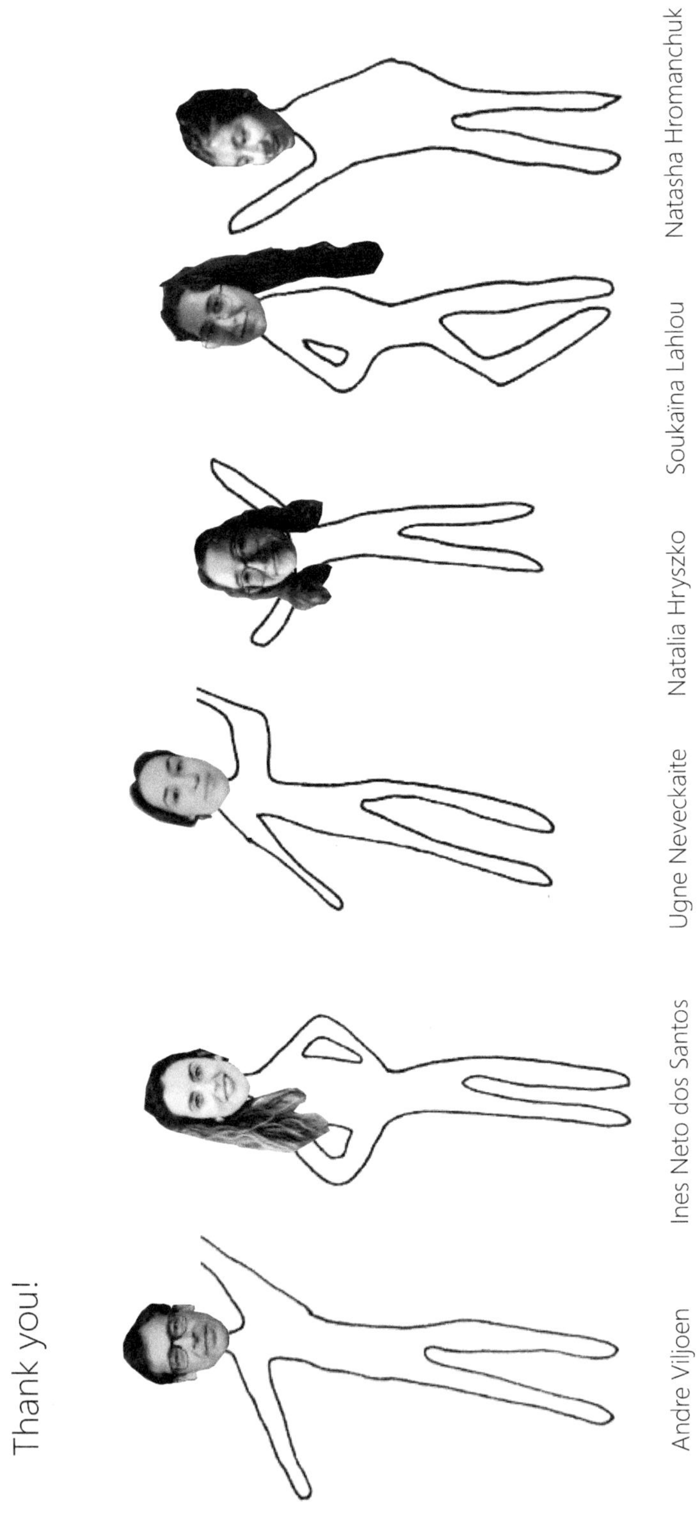

Thank you!
Andre Viljoen
Ines Neto dos Santos
Ugne Neveckaite
Natalia Hryszko
Soukaïna Lahlou
Natasha Hromanchuk

FIGURE 7.1

8

USELESS 3 'HANDS-ON EXPERIMENTATION'

Exploring the potential of the useless

Folke Köbberling and Alexa Kreissl

How can we turn a digital workshop into a practical field of experimentation, venture outside together on a shared mission to uncover the potential hidden within the seemingly useless?

The objective was to collaborate hands-on across distance, overcome the digital realm, and explore discarded materials, resources that we encounter every day and consider waste and useless. Where do we find these materials, where do they appear, and how can we raise our awareness and filter them?

We were spread out in seven different cities in four countries, each governed by varying COVID-19 policies. We had limited tools at our disposal and the concept of the 'Useless'. Our approach involved alternating between theoretical and visual inputs, practical instructions, and exercises to delve into the topic collectively.

The first day started with an introduction into artistic mapping and offered diverse examples of how we can observe, visualize, and trace what we see, experience, feel, and filter. This initial tool was crucial for our journey outside, enabling exploration, mapping, and later sharing and exchanging our findings.

During their wanderings through their respective neighbourhoods, the students developed an understanding of the resources scattered on the streets – the so-called useless. Gradually, they became conscious of discarded goods, discovering more every day.

The useless is ubiquitous, present on the streets, on construction sites, in backyards, in gaps, and then everywhere. Places that we usually ignore and overlook and that take up an enormous amounts of space. Space for waste – a waste of space.

In the following days, various techniques and tools were introduced to facilitate the hands-on exploration of materials, fostering creative solutions and skill development. One method involved modular weaving, a reversible connection technique in which flexible materials can be joined and disassembled. Ring weaving, according to B. Schmeling, describes structures in which closed-looped elements can hold with themselves and be connected to surfaces and volumes. Old bicycle inner tubes and beverage packaging like Tetra Pak can be cut into loops and interwoven. Simple patterns and surfaces or, with practice, very complex shapes can be created by this process. The technique is transferable to various materials and dimensions, even dismantled car tires can be transformed into durable building material.

DOI: 10.4324/9781032665559-13
This chapter has been made available under a CC-BY-NC-ND 4.0 license.

Another method focussed on exploring excavated earth. Excavated earth occurs during construction work and is declared as waste as soon as it is transported from A to B. This excavation, which, at best, contains clay, can be examined using various methods to determine its properties and potential as a building material.

The students went in their immediate vicinity, touched ground, dug for earth, documented the process, and created a map with different soil samples.

The methods of weaving and earth building are deeply rooted in various cultures around the world and are among the oldest cultural practices. These techniques offer the opportunity to experiment with readily available materials at a local level without having to rely on complex tools and additives. Their low-tech nature allows construction to be carried out without external constraints.

During the next stage, participants brainstormed potential applications and material combinations.

Mattress foam was combined with excavated material to create a potential insulating block, and a woven mat made from bicycle inner tubes was layered with a traditional loam mixture as reinforcement.

Despite being a digital summer school, the students were able to engage in hands-on work with various materials. They refined their craft skills to expertly handle a range of materials, dis- and reassemble them, or, as in the case of one student (Shemol Rahman), carry a found piece of furniture across town to the correct location for disposal where no pedestrians were allowed.

The journey was documented in moving images by each participant. The result is a collaborative video that gives an insight into a world that values the potential of the unusable by transforming it into useful materials. Nevertheless, we lacked the common real space that would have enabled spatial cooperation in a larger whole.

By fostering creativity and skill development, the workshop contributed to a greater understanding of sustainable practices and resource optimization. The findings highlight the importance of recognizing the value of seemingly useless materials and promoting their transformation into useful resources. Future research should focus on scaling up these practices and exploring additional innovative approaches to address sustainability challenges.

Earth as found

Raw earth is found in the subsoil of our planet and consists of a mixture of liquid, gaseous, and solid parts, like stone and gravel, sand, silt, and clay. The clay is the binder, the glue in our mixture. Like cement is the binder in concrete. It is found under a 20–40 cm layer of top soil where plants grow and nutrients develop.

There is evidence of raw earth being used in architecture as early as 5000 BC in China, in the form of rammed earth – a process of constructing natural walls and floors using raw materials such as earth, clay, and chalk. Historical examples of rammed earth can still be seen today, for example, on sections along the Great Wall of China.

Raw earth has many advantages and benefits, especially in relation to energy saving, transportation costs, and ecology. Most varieties of soils can be adopted if they include appropriate granular substances: there are a dozen construction techniques, including compressed, moulded, extruded, and poured earth.

Excavation is the soil that is released during construction projects or tunnel borings. Today, this is disposed of at great expense and transported over long distances due to the few suitable landfills. It usually consists of a mixture of loam, clay, sand, grass, and topsoil. Just 100 years ago, every village had a clay pit in which excavated material was stored, and the clay was later used to build half-timbered houses. Even today, excavated material should be understood and used as a resource.

The students took soil samples at various locations in their city, where they found access to the ground. First, they had to remove 20–40 cm of the top layer (Figure 8.1). In the next step, they examined the suitability of the soil as a building material.

For the sedimentation test, the soil is mixed with water and left in a transparent vessel for one day. The different substances are sedimented in layers and show the composition of the soil.

The cohesion test examines the binding ability. For that purpose, you can either send the soil to the laboratory or, like us, perform a field test. After removing the coarse parts from the soil, it is processed with a little water to a pasty compound about the size of an orange. From this, by uniformly molding the ball, a cigar-shaped cylinder is formed, which is about 2–3 cm thick and 30 cm long. Its cohesion is now tested. We let the cigar slide over our hand and see when it stops. If it breaks immediately, the material has poor bonding ability, if it takes 15 cm to break, the adhesive strength is good (Figure 8.2). This must be tested at least twice more with the same material and the results compared. After all fallen pieces have been measured, the average is determined: less than 5 cm pieces, poorly binding sandy soil; between 5 and 15 cm, soil with medium cohesion; and more than 15 cm, a very well binding clay soil.

These tests are very important to determine the properties of the material and the bond strength.

Ring weaving

Ring weaving describes a reversible technique that uses interconnecting flexible closed loops to form flat mats and 3D structures without any further binding substances. They just hold with each other. In addition to conventional joining techniques such as gluing, screwing, welding, and riveting, the process opens up a wide range of possible applications in different dimensions and numerous advantages for sustainable resource utilization.

Each weave remains demountable, and when taken apart, each loop is still a functional element that can be reassembled into a different shape.

The adaptive process is transferable to different materials and formats and can be manually applied on site. It allows to bypass complicated material-specific joining techniques, and various materials can be converted into components of a modular system by forming them into a loop. Materials investigated and suitable for ring weaving come from urban and industrial waste streams.

The German autodidact and inventor Burkhard Schmeling has been researching possible applications for two decades. He has developed a process that breaks down old tires into closed loops with little energy input, thus minimizing their empty volume for storage and transport. The loop, as a new product, is used as a car tire for only a short time in its 1,000-year life cycle and retains the tire's essential properties: tensile strength, resilience, and elasticity. In the weave, it can pave construction roads, be used as a dam for flood control, fix slopes, cover surfaces, and form volumes.

FIGURE 8.1 Moritz digging in the park

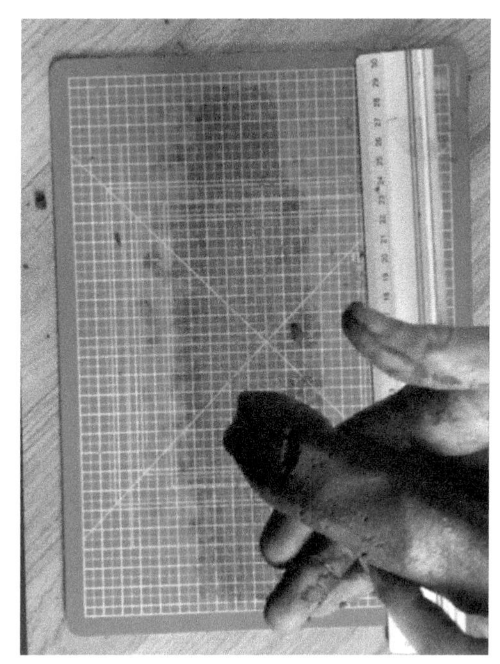

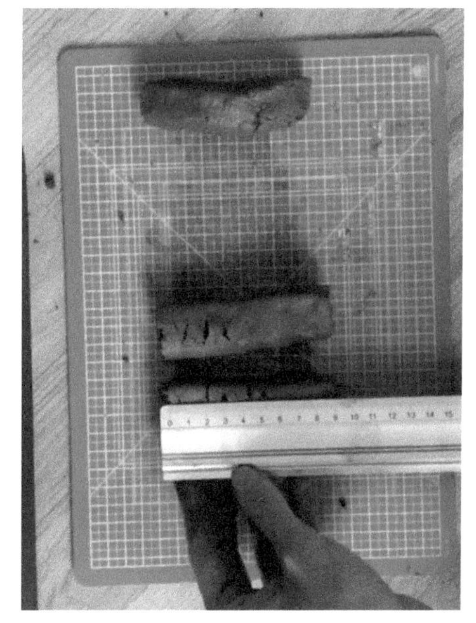

FIGURE 8.2 Cohesion test in Peckham (UK) and Paderborn (DE)

By transforming the adequate material into a closed loop, it becomes a constructive element that can be flexibly joined with other elements of the same shape. Instead of shredding materials with high energy, this method attempts to transform the material as little as necessary, thus saving energy and utilizing the positive inherent properties of the resource, whose life cycle is not nearly fully appreciated today.

Different materials can be joined together in a structure to form material composites, layered into each other, filled and remain separable in terms of type. The cavities between the loop layers can be padded with filler material, causing the joints to expand and the weave to become denser and stronger. In addition, it is possible to insert reinforcing bars through the cavities arranged in rows in the structure to rigidify it in different directions.

Woven flexible joints as connecting links allow straight rods (e.g., wood, aluminum) to be joined together. They can fix or be used as hinges and flexible joints.

As a deconstructible element of a whole, this constructive experiment opens up the resource-saving use of high-quality materials and the repurposing of resources that initially do not appear to be based on any primarily constructive property. Any material that can be formed into loops or coils is, in principle, suitable for the process, and tectonic forces are created in the sum of the individual parts.

This technique enables the design of surfaces and structures of different sizes that can be adapted at any time without irreversibly damaging material. Remaining loops always offer the possibility for new shapes. Small patches can be joined together to form larger ones, and large mats and structures can be separated by intersections and woven back together.

During the School of Re-Construction, the students were introduced to the technique and learned how to apply it to Tetra Pak and bicycle inner tubes, materials common to all of their home countries. This allowed them to apply the technique in pilot testing with the idea in mind that the shapes could also be transferred to larger materials and dimensions, such as car tires.

Used inner tubes can be collected for free at local bike stores. After internal and external cleaning, the tubes as well as the Tetra Pak can be cut into uniform width loops with scissors or cutter and are then suitable for weaving.

Moritz produced a mat whose diagonal deformation he examined (Figure 8.3). He discovered that longer pieces of inner tubes can also be assembled into loops, held in place by the friction

FIGURE 8.3 Material testing

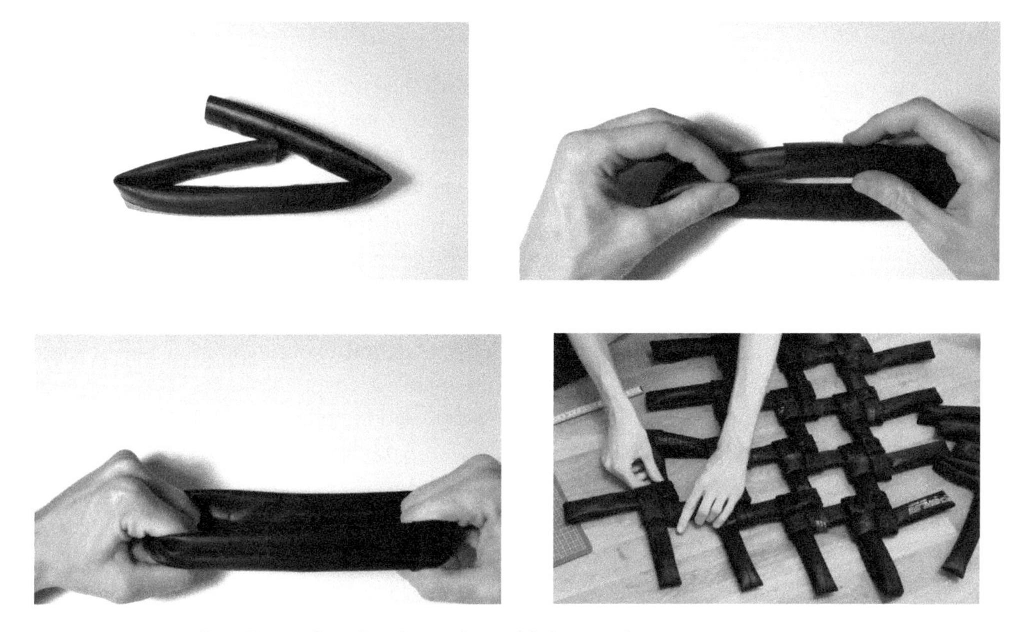

FIGURE 8.4 Loop forming and testing | weaving with larger elements

FIGURE 8.5 Traditional loam mixture: soil with straw and dung | application

and tension of the rubber as it is interwoven. This way, the tubes can also be transformed into larger elements. (Figure 8.4)

Shemol stretched his rubber mat between two supporting rods, used it as a supporting structure and plastered it with a classic mixture of straw, clay and dung (Figure 8.5). Elisabeth explored the possible variety of patterns and shapes and brainstormed about potential applications such as artificial reefs. Despite the long distance, the method could travel and be transferred to local materials and executed by craftsmanship with basic tools at disposal.

Mattresses

Every year, up to 30 million mattresses in Europe reach the end of their life, of which 40% end up in energy recovery and 60% in landfill.[1] As they are considered bulky waste, no consistent data is collected. Mattresses are multimaterial products whose exact composition is mostly unknown to the waste collector, so they can hardly be recycled and continue to be incinerated. Valuable resources are lost and no longer returned to the material cycle. The method is an interdisciplinary loop, but taking a creative approach enables a grasp of the bigger picture by invoking common sense, from aesthetic appreciation through to value creation, which seems to be absent in current commodity production. This includes the construction and waste industry and contemporary architecture, intertwined with architectural experimentation and scientific measurements and testing. With the removal of the mattress covers, different shapes, colours, textures, and foams emerged, whose properties we explored in both sculptural and constructive applications in a participatory manner at the Institute for Architecture-Related Art at the Technical University of Braunschweig. In further investigations, different mattress foams were tested in the climate chamber for their insulating properties, which are eventually equivalent to those of mineral wool.*

For the School of Re-Construction, the students were encouraged to scout the streets for mattresses (Figure 8.6). Moritz experimented with foam and raw earth to build an insulating brick (Figure 8.7), while Elisabeth incorporated them in her idea of artificial reefs.

* Kreissl, A. (2021) "Resource mattress. The potential of refuse materials", AGATHÓN | International Journal of Architecture, Art and Design, 9(online), pp. 184–193. doi: 10.19229/2464-9309/9182021.

FIGURE 8.6 Old mattresses in the street

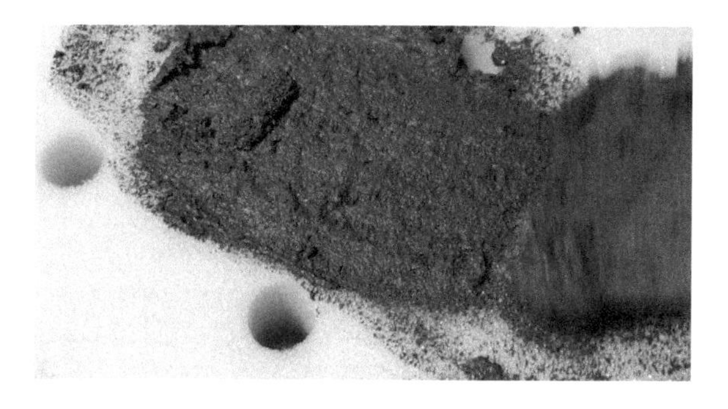

FIGURE 8.7 Mattress foam and earth

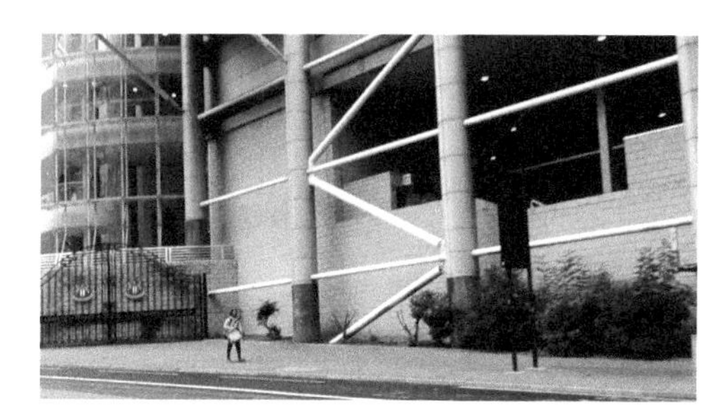

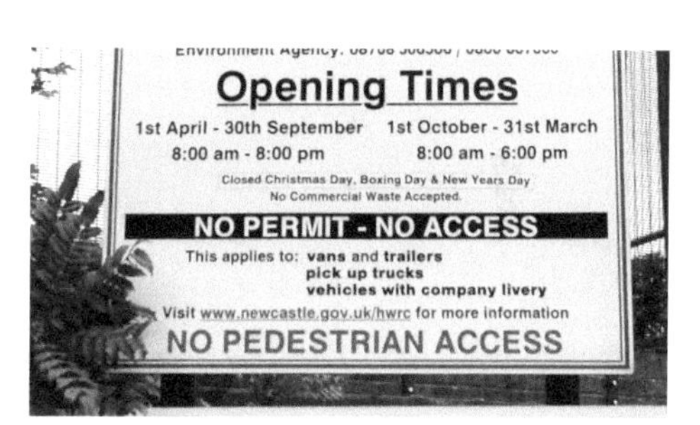

FIGURE 8.8 *Sometimes doing something*, video stills

Source: Shemol Rahman

Sometimes doing something

We were struck by the absurdity of waste found in our neighbourhoods and the space it took up.

Shemol Rahman was reminded of a comment a few of my neighbours made, saying how difficult it was to dispose of large items responsibly without a car, which you need to enter the recycling facility.

Inspired by the artist Francis Alys, who pushed an ice block around Mexico City for six hours until it melted, I decided to expose the absurdity of this situation by walking to the nearest recycling centre with a found item, something someone deemed useless. I wished to critically map the disjointed-ness of infrastructure.

Walking in the footsteps of Sisyphus, I wished to show that we may not necessarily be doomed to forever fight an uphill battle with our waste, instead, that finding the usefulness in the useless, finding meaning in absurdity, is reason and purpose enough to carry on and, sometimes, do something.

In hindsight, pushing a sofa or fridge around would have been more absurd. The point isn't to encourage people to throw away more chairs, but to build local, accessible re-use infrastructure.

Shemol Rahman

Note

1 Fraunhofer ICT (2019), "Was passiert mit 30 Millionen Matratzen pro Jahr?". ict.fraunhofer.de.

9

BY-PRODUCT 'METHANE'

Michael Howe and James McAdam

An unreliable historic background to our Methane Team brief

The history of new materials and techniques, developed from the discarded elements in any production process, is long. The ideal intensively managed agricultural (year) cycle gives clues to the benefits which might be accrued. For example, take the humble pig as the centre of traditional Chinese and Northern European husbandry, combining domestic sanitation, agricultural production surplus waste management and animal rearing. Through the pig's diet, household organic waste produced from late spring through to autumn is transformed into fat and protein for the winter, (with the helpful by-product of fertilizer for the plants which feed the family). This is, we suggest, a circular diagram to be studied and emulated in the building industry.

A story which describes a similar crossover of interest streams was told to me some years ago (the venue for the tale was a bar in Chicago, so I make no claims for its veracity, but it sounds right). The Owens Corporation (USA), an early manufacturer of insulating glass fibre is said to have commenced development of this product as a result of a "waste management" problem associated with the production of soda bottles.

Apparently, each counter blow during the injection-moulding process developed a small filament of glass when the die was withdrawn. Hard to compress and impervious to burning on fires, (the individual surface resistance of each element in combination proved too good at thermal insulation), the material would only succumb to remelting in large and costly furnaces. Burial was haphazard as the light "wool" would blow away before it could be cast into the pit and covered.

Something had to be done. Luckily, the U.S. government was looking for very light weight insulation for high-level aircraft designs on the drawing board prior to WW2. The rest is aviation history and, of course, building industry history, or so I have been told.

On safer historic ground, the shifting relationship between human waste and production is well-attested. In the past, our ancestors very sensibly used urine as a vital component in the preparation of wool, textiles and tanning skins. While this may have been a little on the malodorous side as a productive application of our waste, it is surely better than just flushing it to very expensive water treatment centres.

DOI: 10.4324/9781032665559-14
This chapter has been made available under a CC-BY-NC-ND 4.0 license.

With this idea of the relationship between waste and production from differing streams, we proposed a small study of methane. The "greenhouse gas on steroids" as I saw it described in *The Economist* some time ago. Rice-producing wetlands, natural bogs and, of course, milk and meat production (we will leave the petrochemical and gas pumping industry aside, perhaps?) were areas for consideration.

Team C: members background

We are architects and designers. We are not scientists. Our work consists of propositions based on the imaginative synthesis of contextual conditions. The nature of the School of Re-Construction encouraged a fast and nimble approach to both contextual information harvest and proposal development.

Team C was made up of a small group of team leaders/tutors, including three practitioners who teach at the University of Brighton architecture department and four international delegate members who at the time were second-year degree students studying at various European universities but whose "home" was, in some cases, beyond the borders of Europe.

Team C members and background

Dorianne Dupre – École Nationale Supérieure d'Architecture de Paris-Belleville, France
Vinciane Gaudissart – University College Dublin, Ireland
Yagmur Gur – Eindhoven University of Technology, Netherlands
Murjanah Uwais – University of Nottingham, UK

Technical and computational team tutor members

Matthew Walker – a senior associate at WOO architects, responsible for the design and delivery of international sports and public buildings. In addition to being an architect of repute, Matthew is an experienced digital design practitioner.

James McAdam – a designer/maker with a specialist interest in digital fabrication and traditional handcraft skills. James studied design products at the Royal College of Art, UK, obtaining his MA in 2002. He has since designed exhibitions, retail and visual marketing, architectural scale art installations, interactive book publications, pyrotechnic products and digital/craft solutions.

Team coordinator

Michael Howe – one of the funding partners of Mae Architects (currently a Sterling Prize nominee), he was responsible for the production of government policy documentation and urban scale housing and public realm design. Having retired from practice, he is a full-time senior lecturer at Brighton University, UK, for MArch technology, undergraduate design studio and intermediate year professional experience. At time of writing, he is a RIBA accreditation assessor.

Team C: methodology

As has been stated earlier in this publication, the digital School of Re-Construction summer school ran from Monday, 2 August 2021 for two weeks, coinciding with one of the UK's COVID-induced lockdown periods. As a result, all meetings, presentations, workshops and tutorial discussions were held on Microsoft Teams with production outputs being shared via Miro Board.

The short duration of the summer school implied a fast, "durty" and fun approach to both investigative method and proposal development. All proposals were to take the form of "provocation" rather than methane by-product solution.

Following our first team "get to know you" meeting, work commenced with three short introductory area background presentations.

The first of these presentation, titled "A Small Planet With Limited Resources", consisted of a brief reiteration of key issues of energy use and the contribution of fossil fuel and organic combustion as it bears on climate change, moments of paradigm shift. While many team members were aware of the issues at play, (they would not have signed up for the summer school if they were not), it was felt that a brief presentation would assist in galvanising a common ground for future conversations.

The second presentation took the form of a description of methane – chemical make-up, production sources both natural and man-made, dissipation and potential uses. This included a description of sites containing visual and scientific monitoring information such as National Aeronautics and Space Administration (NASA), and Greenhouse Gas Emissions Monitoring Service (GHGSat), useful for baseline research.

The third short presentation consisted of a brief introduction the teaching practice of the team leads including reference to the student-initiated movement "Architectural Education Declares". Past Brighton MArch student digital analysis of landscape and building forces and environmental conditions as examples of "tool-based investigation" were discussed. In addition, an example of a previous research project undertaken by tutors and students from University of Brighton and Oxford Brookes that uses "Waste stone" to create utile masonry structures as a potential vehicle for social and economic regeneration was described.

Our means of communicating our ideas and proposals for the summer school formed a key point of departure as the team moved into the next two phases of operations, information harvest and digital skills development. A statement of communicative intent was proposed.

Team C: communication credo

We live on a Small Planet with Limited Resources
Humans still want stuff, and We design stuff, so We are going to have to do more with less.
In order to avoid monomaniacal system design, (associated with first and second wave Industrialization), more voices should be heard.
We believe that the Best Decisions are made by the largest workable groups.
That means consulting apparently "non-expert" as well as "expert" groups.
In order that we communicate proposals well to disparate audiences our information should be presented in an aligned manner.

What do we mean by aligned information?

It is simple really. Never use a diagram, which requires expert knowledge to decode, when a simple model or animation (the product of powerful but widely available computation capacity) can give a clear quantitative and qualitative visual description of conditions. With this approach to our output communicative capacity, the brief was introduced.

The Team C brief was framed in three parts and put to the delegates in the following manner.

A. We would like you to consider methane in your neck of the woods and to compare the volume and sources of methane with your fellow student delegates. In order to do that, we wish you to produce a terrain and methane model covering a 15 km (east/west) by 10 km (north/south).
B. Once we have these data, we can start to identify the source/sources of this gas in each area.
C. When the source is identified, we will make sketch proposals for harvesting and using all this free bounty.

It is a testament to the delegate's perspicacity that they did not run when they realised that we expected them to grapple with new, to many of them at least, software for modelling. Matthew (who was just about to go on paternity leave so he was very busy at the time) produced a number of modelling and landscape mapping tutorials that he prerecorded for delegate instruction (just in case his first child appeared during the summer school). These tutorials included sourcing satellite imagery from both NASA and PULS of methane "hotspots" and topological photography so that the delegates could produce their own regional methane emission maps. He maintained a constant email Q+A with delegates to help refine their work and answer any software or technical issues. (All software used was part of a Rhinoceros 90-day free evaluation so no costs were incurred by the delegates).

The outcome landscape models proved invaluable when the team attempted to identify and communicate possible methane sources locally. During group tutorials and workshops with James and Michael, these sources were interrogated and reactive approaches discussed and developed by the delegates.

It was noted by one of the delegates, Vinciane, when investigating her chosen area of the Dublin Bay area, that while methane produced in the city and suburbs increased somewhat during winter months, due to increased domestic heating and possible car use, this was not as significant as the year-round "hotspot" identified in the coastal area southeast of the city.

Situated as a coastal strip including Bray and Shankill, it appeared to be a genuine result of natural coastal emission. (While this may have some link with the gas terminal at Bray, it was decided by the team that the area affected is a little too extensive to be as a result of this factor alone).

Vinciane argued that the "hotspot" situated in a wetland area (wetlands contributing by organic breakdown to methane production) is bounded by relatively shallow coastal water. This location may also contribute via natural seepage from the ground, a possible source being small

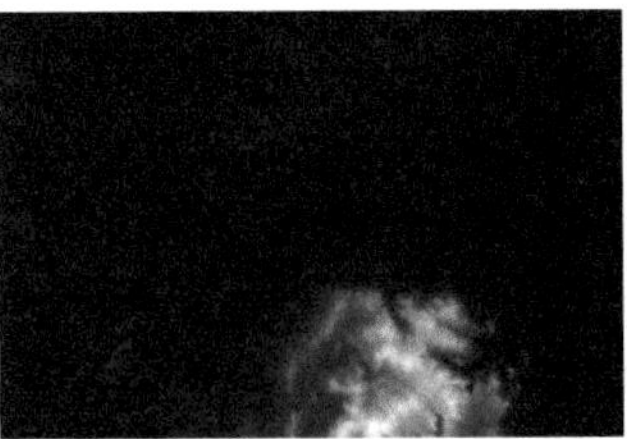
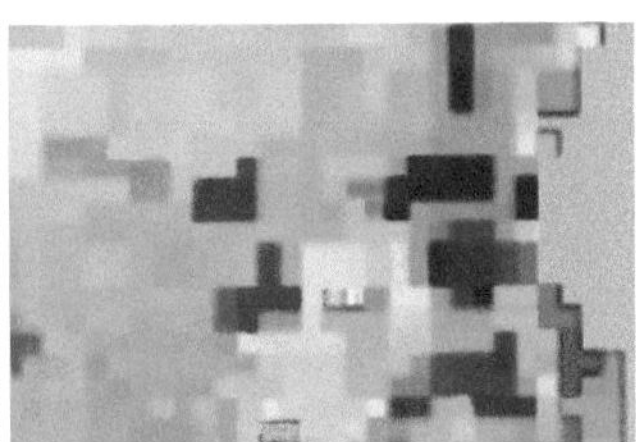

FIGURE 9.1 Methane in My Back Yard: Desktop identification of year-round consistent high methane sources in the South Dublin Bay area

Source: Delegate Vinciane Gaudissart using PULS data from 2020

FIGURE 9.2 Commencement of Vinciane's methane terrain map. The XY axis has been manipulated to present high methane areas as taller features in the model landscape.

natural gas reservoirs or more likely produced in oxygen-depleted sediments on the sea floor. The process of seaweed and marine plant life rotting or, indeed, the product of phytoplankton are possible contributing factors here.

While Vinciane could continue her work in her adopted city of Dublin during the summer school and lockdown period (having previously attended the European School of Luxembourg for 14 years), another of the delegates deserves special recognition of her working situation at this time.

Delegate Murjanah Uwais, a student of product design and manufacturing at the University of Nottingham in the UK, was subject to travel restrictions in place at the time, which left her unable to leave her "family home" of Abuja, Nigeria, for the duration of the summer school. Contending with sporadic power cuts and variable internet availability, Murjanah proved to be a robust contributor to team discussions of local methane environments when she could connect with team meetings.

Her methane terrain maps prove some of the most dramatic examples of manmade (agricultural and petrochemical usage) methane event variations for the year 2021.

FIGURE 9.3 Murjanah Uwais "Methane in My Back Yard" July and February methane terrains for Abuja, Nigeria, 2021.

The upper model (July) shows methane production mainly confined to urban conurbations, while the central map (February) shows methane production as universally high across the area under investigation, including both urban and agricultural land.

Comparisons with the European area models proved enlightening to all team members. For while the European maps tended to show fields dominated by green areas with moments of high (red), methane production, the Abuja map for the "hot period of year" (March being the hottest month with averages of 90 °F) shows an almost unbroken field of high methane event. Red areas denote methane concentration between 1,930 and 1,980 ppb. Green areas showing between 1,710 to 1,800 ppb approximately.

In the opinion of Murjanah, during the relatively cool period of the year, July to August, methane production can broadly be attributed to vehicle traffic, cooking, and electricity production by petrol and diesel generators (for off-grid areas or during variable power supply from the grid). The almost universal high levels her model show during the hottest time of year, January to March, while still maintaining the previous contributory factors should be attributed to agricultural practice, that is, fire, intended and unintended, as there is little or no rainfall for December and January.

Another factor, which is, perhaps, not so critical in the area around Abuja, is leakage from oil and gas production infrastructure. Nigeria is a member of OPEC, and its production in this field is viewed a major contributor to the national energy security and governmental revenue generation. However, the infrastructure and pipe network associated is subject to the twin stresses of intermittent maintenance regimens and illegal interception or tapping of these resources. Both situations contribute to high levels of methane seepage and land and water pollution in some areas.

Delegate Yagmur Gur of Eindhoven University of Technology, Netherlands, chose as her "backyard" an area of Holland she was familiar with, Zeeland, an area with diverse potential sources of industrial and agricultural (including cattle and dairy) methane production. The site of investigation also includes the Zeeland Oil Refinery. The refinery shows as a consistent high, (red) spot on the PULS information included in the desktop study. Some variation in general

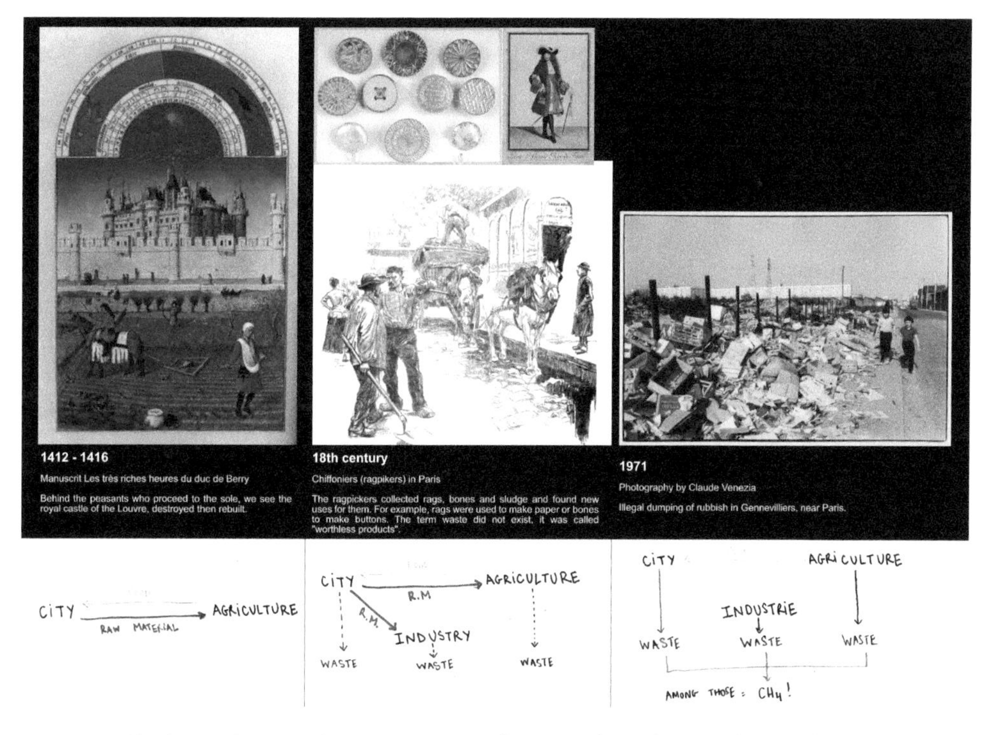

FIGURE 9.4 Dorianne Dupre: Commencement of proposal. Paris: A relationship of agriculture, industry and waste.

background methane was attributed to increased domestic heating in winter but also as a result of agricultural practice due to the increased use of silage with its associated rotting and cattle flatulence.

The Team C Paris delegate, Dorianne Dupre of École Nationale Supérieure d'Architecture de Paris-Belleville, chose to take an optimistic view of the increased potential of food productive urban landscapes. The increase in these social endeavours, which deliver benefits of personal psychological well-being and social interaction, fresh and good quality food at point of consumption and reduced family food bills with reduced carbon required for haulage, bring with them the potential of increased methane production in urban areas due to the action of composting and organic material breakdown.

Dorianne's proposal consisted of relatively small, highly "local" methane harvesting infrastructure to be initiated with all new urban food production sites consisting of fermentation tanks for production of composted fertilizer. The harvested methane would provide a fuel source for heating growing environments on site increasing the variety of plants available to this North European city and extending the "growing year".

Simple, classy and "do-able", we thought.

University College Dublin delegate Vinciane Gaudissart's proposal also presents a small scale, local "methane as fuel application" delivering a big social punch.

Developed, in part, as a result of her twin interest in the "housing crisis" faced by young people in many European cities, including Dublin, and her interest in low-carbon building solutions, as witnessed by her involvement with Architects Climate Action Network (ACAN) Ireland, the proposal is designed with a view to humane neighbourhood regeneration for the city and suburbs of Dublin while maintaining the embodied carbon of existing buildings.

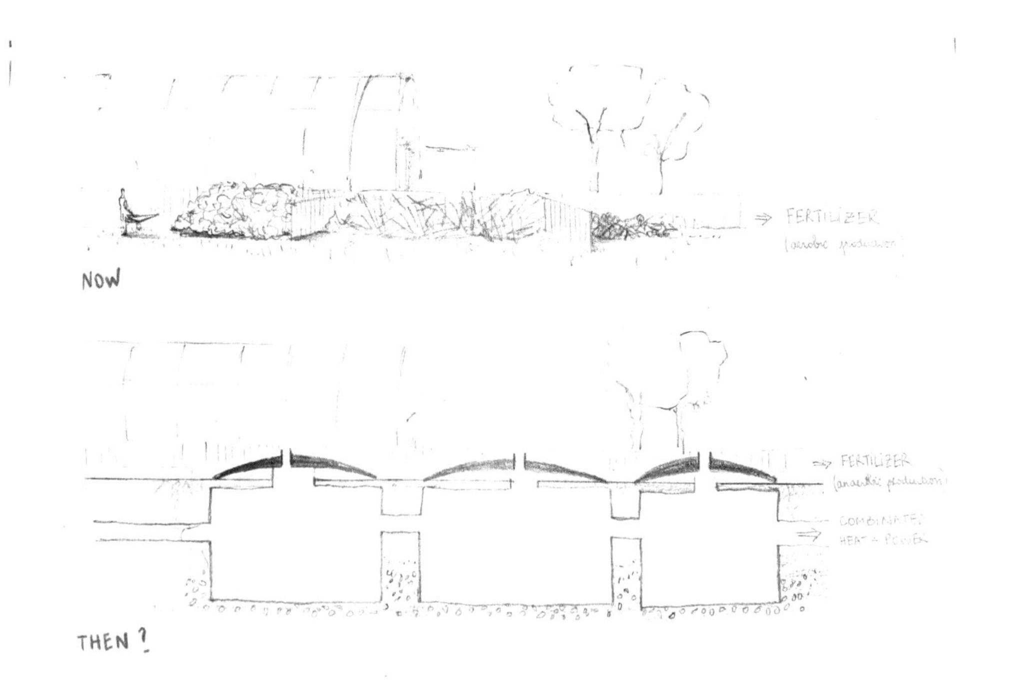

FIGURE 9.5 Dorianne Dupre, Methane proposal: Allotment waste = Fertilizer and Power.

The delivery end of the proposal consists of the "twinning" of a young person or couple who do not have a home with a run-down or dilapidated property. By utilising "sweat equity", the young person can take responsibility, with training and support, for the renovation of the property. The "sweat equity" developed during renovations allows for a reduced (affordable) rent.

One of the key issues identified by the delegate was the period before the property is occupiable, due to a lack of service connections and basic sanitation and cooking facilities, when the young person must be housed and the ground rent and other expenses associated with the property for renovation must be met. Her proposed vehicle for early occupancy is the installation of temporary, pop-up, hot water and cooking facilities all powered by locally harvested methane.

Illustrated in Figure 9.6, the methane service pod is delivered so that the property is now a relatively warm and clean "camping site" during the building skills training period. The final

FIGURE 9.6 Vinciane Gaudissart – Homesteading in Dublin. a. Rundown property twined with homeless young person. b. Methane-assisted camping period for training. c. Sweat equity = Fixed building, (low-carbon solution) + homed person, (with affordable rent).

FIGURE 9.6 (Continued)

image represents the resident renovation period, at which time new services connections mean that the methane pod can be removed and established in a new property for renovation.

In order to facilitate the methane housing renovation support system, methane must be harvested. The proposal for this consists of a meandering wetland shore trail, connecting various public access recreational structures, such as boat moorings, fishing and swimming decks, refreshment venues, etc. Proposed as the South Dublin Bay Park, the structures will form methane harvest points along the route from South Dublin to Shankill.

This is a brief description of some of the activities undertaken by Team C as not all work has been presented. It is our hope that some idea of our time with the School of Re-Construction is given. All in all, it was a most entertaining and enlightening two weeks in the middle of a less then fun period of lockdown.

Michael Howe 2022

FIGURE 9.7 Proposal for the South Dublin Bay Park. Public access recreational structures connecting a seashore and wetland walkway. The access structures act as points of methane harvest.

10

HYBRID: COMPOSITE/RE-USE/RE-MIX

[The dub tracks]

Katarzyna Sołtysiak and Anthony Roberts

> *We can re-make parts of the city without producing new construction materials while at the same time creating another idea for beauty – a beauty that rises out of the act of re-assessing what is called‘useless’| ’discarded’ and generally considered ‘worthless.’*
>
> *Anthony Robert's*

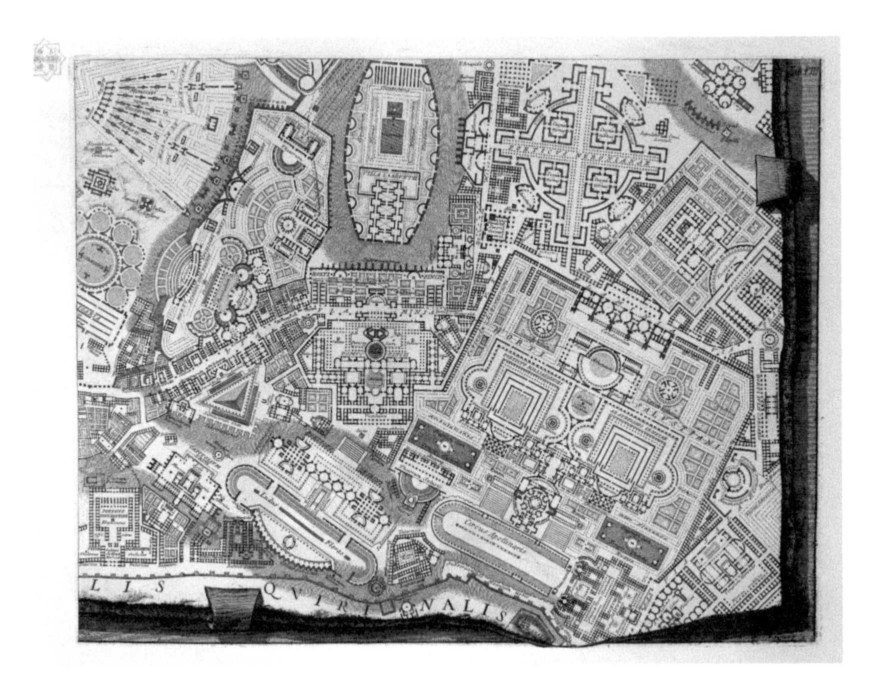

FIGURE 10.1 Robert Adam, Giovanni Battista Piranesi, ‘Map of Campus Martius’, 1762

DOI: 10.4324/9781032665559-15

This chapter has been made available under a CC-BY-NC-ND 4.0 license.

Initially, the word 'hybrid' meant (the)'cross-breeding' of two different species. With time, it could mean more distinctively different species/elements combined. What is characteristic of a hybrid is its 'mixed character'.[1]

For us 'hybrid', in terms of architecture can be applied on many levels; using 'found' materials and reconfiguring them to create something new. Secondly, a city itself should remain a montage of multiple parts that encourage diversity, as Rowe and Koetter state in Collage City.[2] Finally, in the era of the Anthropocene, the entire planet has become a hybrid of pre-human tissue and technology, and we all have an impact on its shape so it is important to make it a positive one.

The task of writing this chapter came to me very unexpectedly and relatively late in the process of composing this book. Initially, Tony Roberts, a senior lecturer at the University of Brighton, was to execute the chapter – as the idea for the brief came from him. My input was only to include a brief reflection on the leading theme. Eventually, this text became a hybrid in itself: the chapter incorporates Tony's notes and documentation, sometimes including entire paragraphs of his.

Conditions

Over the course of two weeks, we worked as a diverse group: varying in age (from 21 to 84), education (architecture, engineering, heritage studies) and cultural background (Chinese, Middle Eastern, European). While some students were at the beginning of their academic life, one of them held a PhD. This resulted in our goals and interests varying, from the experience of a journey to hope for a standard architectural project; as tutors, we wanted to avoid the latter as much as possible and focus on the process instead. We were hoping to provide the participants with a set of tools and a platform to exchange ideas and observations. Hence, at the beginning of organising the workshop, the question appeared:

:how can we arrange participants from diverse places and different experiences so as to create a series of projects that 'taste' and have the 'fragrance' of a coherent multiple collectivity/an adjacency of ideas, discrepancies, and conjunctions?

Goals

:what needs to happen is to identify, classify, and re-assign the existing 'waste' materials to projects in those places where we can re-make parts of the city without producing new construction materials: at the same time we need to create another idea that deals with 'beauty' – a beauty that rises out of the re-assessment of what is often called 'useless' or 'without value'[3]: it is time that ethics replaces aesthetics.

With this approach in mind, cities can be perceived as material banks where each new construction appropriates the existing. No material is simply landfilled. This, however, requires new approaches, stepping outside of a typical professional framework and imagining new futures. Continuing the practice of erasure and insertion must end. The first step, perhaps, is to start perceiving cities as material banks and change the idea of mining. We cannot continue to extract virgin materials to construct new buildings – we must start looking around and utilise what is already here.

Approach

: *in this'zoom' age our geography has, in effect, become more-or-less redundant and yet, there is a need for 'super-specific' contexts to become implicit in the projects developed in this summer school.*

Initially, this workshop was not supposed to take place online – Brighton Council prepared materials and space for the purpose. When teaching about materials, a physical contact with such seems crucial. In professional practice, there is a lot of distant treatment: surveying sites by Google Maps, assumptions made on precedents, etc. For this reason, we encouraged students to inspect their immediate surroundings within a given range. Later in the process, these local conditions were collaged together. The whole process was to resemble a game. We encouraged a less specialised, more 'amateur' approach.

: *by utilising a revised model of the innocent game, 'le cadavre-exquis', we will research, re-build + re-invent parts of our collective town: these 'new' hybrid spaces will be drawn and modelled and then (digitally) composed from the local, 'borrowed' and 'stolen' materials sourced from each of the administrative districts where the six 'urban' events are located.*

: *this 'architecture of re-invention' is a game with a serious purpose; it is intended to both inform and encourage a more 'sensible' and 'sensate' model for our architectural and building practice that genuinely attempts to honour the ecology of our planet.*

'Le cadavre-exquis' or 'exquisite corpse' is a collaborative form of creation invented by a group of surrealists. By liberating from context and increasing the role of chance, it encourages the creativity of the participants. It results in the unexpected and juxtaposes parts that don't belong together. Here is a game description from *A Book Of Surrealist Games*:

> **For three or more players.**
>
> Each player receives a sheet of paper and folds it into equal sections, as many as there are players, and usually with the lines horizontal to the proposed picture. The sheets are smoothed out and each player draws whatever he will in the top section, allowing the lines to cross the crease by a few millimetres. The sheet is then refolded back onto this crease to conceal the drawing and passed to the next player who begins the next section from these lines. And so on, until the last section, when it is unfolded and the result revealed. (The sheet may be passed back for the first player to furnish it with a title before the picture is revealed).[4]

Because of the collaborative nature of the game, nobody can claim ownership of the exquisite corpse. As the drawing is composed of the work of several individuals, it shows formal diversity and prioritises process over an origin and the outcome. Similarly, in the organic growth of a healthy city, nobody can claim a singular ownership over its creation.

Collage City

The game (Exquisite Corpse), just like collage or montage, stems from idea of juxtaposing elements in order to grant them a new meaning and explore unexpected relationships between them. Before the term 'collage' came into widespread use, artists experimented with techniques involving the layering and juxtaposition of different materials. The Dada movement, in particular, played a pivotal role in popularising this method, placing a strong emphasis on the use of found and everyday objects, elevating them to the status of artworks within the context of their creations.

Simultaneously, surrealists found collage intriguing due to its exploration of chance in the creation of art. The practice of juxtaposition evolved into artistic games, most notably the Exquisite Corpse. Initially applied to textual materials to create a unique form of poetry, it later transitioned into the realm of visual art. The core idea behind this game was to stimulate the imagination and collaboratively forge something entirely new.

Mies van der Rohe embraced collage techniques from the surrealists, transplanting them into the realm of architecture where they have remained a standard representation method ever since. This shift marked the beginning of architectural representation surpassing the actual architectural product – an influence still discernible in contemporary architectural renders. However, there existed an alternative approach rooted in the art of collaging.

In the 1950s, a group at the Austin School of Architecture, known as the Texas Rangers (comprising John Hejduk, Colin Rowe, Robert Slutzky and Bernhard Hoesli), initiated a unique exercise called the 'plan game';

> During the intense heat [of Texas] Colin, Bernhard, Bob and I played a game. I think Colin and Bernhard invented it. We would take a large blank sheet of drawing paper and begin to draw plans of buildings, historic and otherwise. Colin would say I am going to draw the plan of the Villa Madama, then Bernhard would draw the plan of Wright's Gage House, etc. . . . All night long, in the early hours of the morning the paper would be filled with plans from all times, many hybrids too. At the end Colin would be devilishly amused and delighted. In retrospect, who would have thought those plans of Classicism, Neo-Classicism, Modern Constructivism, [and] Contemporary would have been the genetic coding of the architectural monsters which followed?[5]

Their creative process was a fluid one, with each participant contributing plans, resulting in a multitude of hybrids. This exercise served as a critique of the prevailing notion of total design, a concept widely favoured by modernists of their time. The Texas Rangers viewed a city as a collage of smaller communities and diverse uses, with their drawings serving as a manifestation of this perspective. Perhaps, their biggest inspiration was work of Piranesi, who pioneered hybridising information. In his work on Campo Marzio, Piranesi intricately weaved together ancient ruins, classical temples, grand arches and novel structural forms in unconventional combinations through meticulous research and artistic imagination.

This concept of the game was further refined by Colin Rowe and Koetter in their book *Collage City*. A key concept of this writing was rejection of total design in favour of more fragmented one. Written at the time of flourishing modernist planning with its social ideals, they criticised it as analogical to the creation of Utopia. Utopia, according to the authors, is not only not attainable but also does not offer choice – an elemental value of a free society. They cited Disneyland and the works of Superstudio, a radical Italian visionary group of the 1960s, as examples where idealism had ventured into extremes, rendering them impractical for real-world applications. Instead, Rowe and Koetter proposed an approach centred on fragmentation, 'bricolage' and the transformation of existing urban fabric to create 'pockets' of utopia within the urban context.

Bricolage is a term that originated in the field of anthropology and was later adopted in various other disciplines, meaning making the best of what is available in the immediate surroundings rather than relying on specialised tools or materials. It often implies a willingness to improvise and a rejection of traditional or standardised approaches in favour of more flexible and context-specific solutions. In the case of city planning, this would mean maximising the potential of a site with all the locally available resources. By reconfiguring these in a creative, manner, one creates a more democratic and diverse city. In such a scenario, an architect stops being a judge of what is good and what is bad and embraces what is actually present.

How does this approach reflect on our theme? It not only implies diversity within a unity characteristic to a hybrid but also calls for hybridising historical value (also in material terms) with innovative thinking: 'Could this ideal city not explicitly function as both a theatre of prophecy and a theatre of memory?'[6] In this context, the ideal city thrives on juxtapositions, offering its citizens a rich tapestry of variety – an essential characteristic of a liveable city as advocated by many who experience the pitfall of modernity, e.g., Jane Jacobs. While many cities tend to evolve organically in such ways, Rowe and Koetter stress the importance of strategically integrating new designs within the historical urban fabric. Total designs of complete erasure and rebuilding have proven to fail on numerous occasions. Social estates in the UK may be one grim example of this procedure – in futile attempts to remove social problems related to crime and poverty, local authorities would demolish and subject the areas to a complete 'renewal'. Let Elephant and Castle in London be a relatively recent, unfortunate proof.

Brief

: the brief is an experiment in a form of a game; our thesis and associated testing will create a toolbox that will (theoretically, at this stage) re-build a small section of each participants' towns without any erasure of the existing fabric: this exercise will document and re-distribute re-usable materials found in the immediate context in a re-mix within the 'patchwork' urbanism we are to establish.

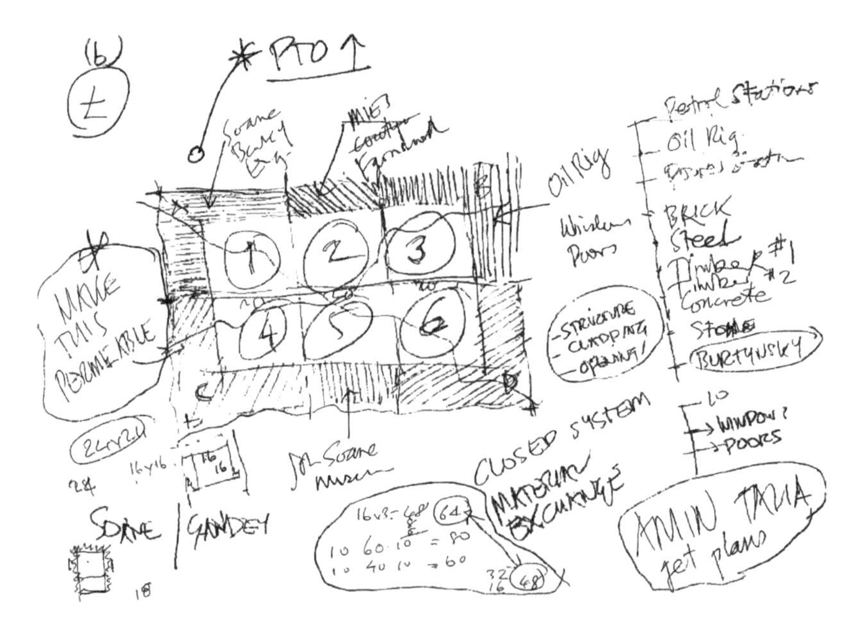

FIGURE 10.2 Anthony Roberts, brief brainstorming

: *each person selects a local piece of their town/city (100 metres x 100 metres square to begin with).*

: *these individual 'territories' form a digital matrix that becomes a new composite – a collective terrain: the aim is to 'construct' a small, collective set of spaces that represent each students' physical territory (and way of thinking) that would also allow differences to influence the neighbouring sites: individual territories are joined together to create a new geography and field of action: this new geography serves as a collective material bank for careful interventions.*

To begin with, each participant defines an area of 100 m by 100 m to explore and 'mine' for materials. Ideally, this chosen site contains both 'vacancy' (to be investigated individually) and a range of materials. We start with collaging these areas over a matrix of six tiles, side by side. It is important to keep the collage updated to share information with other participants.

The following day, a discussion takes place: What sort of events occur in the area and what materials are available? Together with other students and team leaders, each person defines a strategy for documenting their existing stock. This documentation initiates a common 'material bank': a list of materials available across the sites. Which materials dominate across this new combined geography? What is scarce? The process of documentation may take several days as it requires physical engagement with each area of investigation as well as creating an inventory; the latter can consist of 3D models, sketches, photographs or a written description to mention just a few.

While participants work on their inventories, team leaders present 'intruders'. These are carefully chosen projects from architectural history; they supply 'ethical' materials and bring another layer to the narration. They also create an additional, 'fictitious' part of the 'site' and are intended to stretch the idea that, in fact, every site can have the 'qualities' of being 'exotic' and 'exciting' if we take the chance to treat them as such.

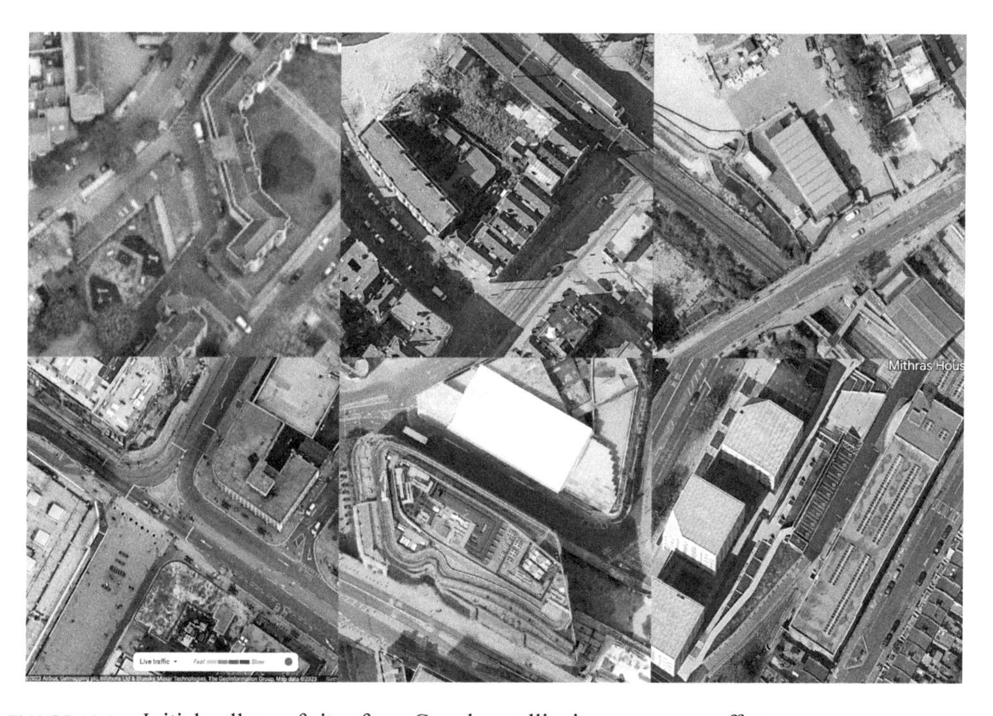

FIGURE 10.3 Initial collage of sites from Google satellite images, group effort

EXISTING MARK- staicase on blank wall

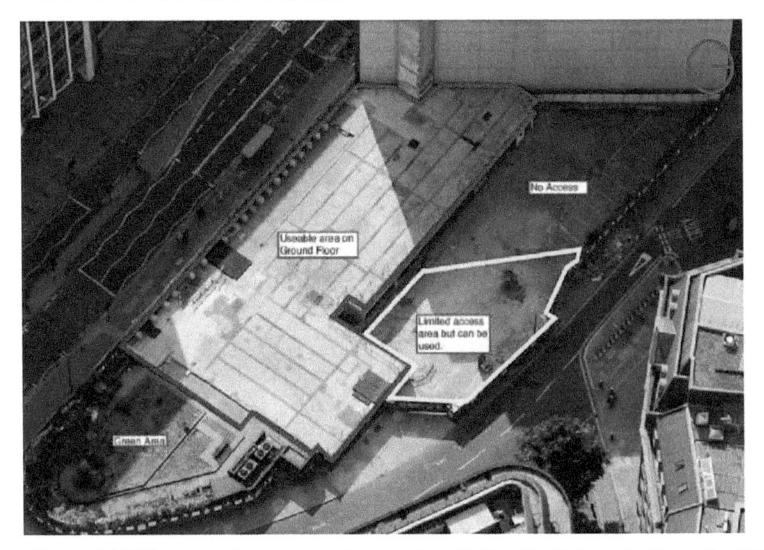
No Access
Useable area on Ground Floor
Limited access area but can be used.
Network Rail have the rights to a natural air Ventilation Shaft from the Tunnel, which provides free air movement into the tunnel . The Garage Extract Plant at Lower Ground Floor also discharges into the Ventilation Shaft, through the sideof the Shaft

FIGURE 10.4 Aidana Roberts, site analysis

By the end of the first week, the collective material bank is complete. Students know what resources are available to them. Now, the politics of 'bricolage' begin to rule: we utilise whatever is at hand and this becomes a base for each individual project.

The second week is devoted to the creation of a project. Each site is narrowed down to another field, this time of 20 m by 20 m. We suggest small interventions – urban acupuncture. No erasure is allowed although students can envision reconfiguration of the existing tissue (especially, when a building is disused and condemned to demolition in the near future). It is important to keep in mind the limited time span – for this reason, we remain in the conceptual stage. It is, however, important to consider the potential of joining the elements derived from various sources.

The workshop concludes with a presentation of each project and a collective drawing of an imaginary collectivity. Nonetheless, the process remains the main focus and the outcomes are only provisional in this brief.

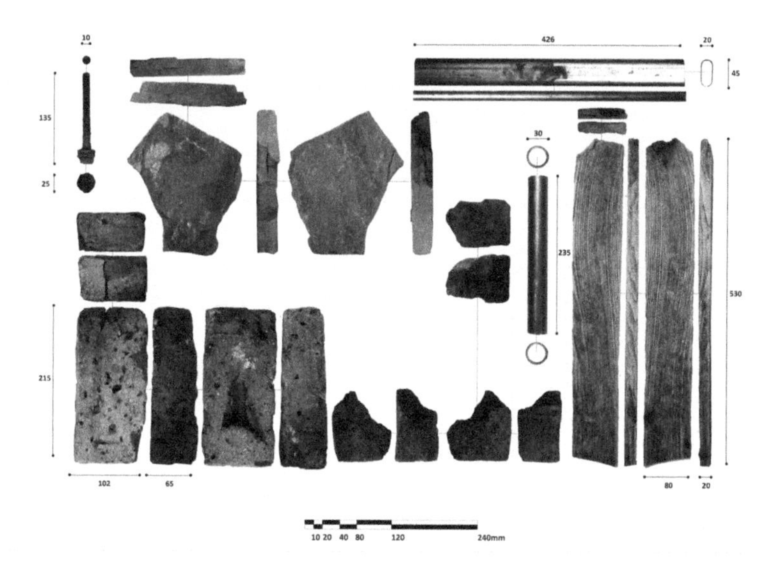

FIGURE 10.5 Weronika Walasz, archeology of Brompton – collage of found materials

Process

Our hope is that gamification of the experience brings an element of fun to the learning process. At the same time, the very serious issue can be presented in a more approachable manner, acknowledging diverse fields of study and prior learning experiences. By engaging in (re)creation of a collage city, one becomes a user and a creator rejecting the modernist idea of 'tabula rasa' and embracing the diversity existing only in dense urban contexts. Such an approach democratises city-making.

Nonetheless, the combination of in situ explorations and online lectures required significant engagement from all participants. The tight schedule would not allow it to be followed alongside full-time work or studies, which caused some to drop out.

None of us met outside of the virtual context for the purpose of this workshop; nonetheless, we were blessed by largely sharing the GMT time zone, with a maximum two-hour difference. While, for some, this was a blessing – one could combine this experience with work or travel – for others, it proved to be very limiting. Without a doubt – the online format hindered a sense of community and collaboration. Students presented their own stories and listened to the others, creating a unity of work that felt rather artificially steered by the team leaders. On the other side, the hybrid nature of this exercise allowed us to connect geographies which otherwise would not coexist – ironically, five out of six were located in the UK, including three sites in London. Yet – even this potential homogeneity resulted in a diversity of conditions.

The initial part of the process was met with significant excitement: I believe one reason for it was that it involved an offline activity (in the pandemic context) and they were all excited to share their choices. Their judgements were made intuitively as the students were not necessarily familiar with salvaging materials. In some cases, temporary elements (like scaffolding and skips) or pieces of functioning infrastructure (parking areas) were confused with salvageable ones. Furthermore, to give the findings any sort of order required introducing drawing and documenting techniques by team leaders. Discussions regarding salvaging the existing stock and its documentation spread over Week 2 and prompted a shorter project phase. For this reason, students could have lacked conclusions and felt like they missed outcomes to present. With the emphasis on the process, we did not see this as a problem – in the end, the learning experience was successful.

Outcomes

The projects that emerged from the workshop included a city gate, a small shelter for the homeless and a pedestrian bridge. Each of these designs was initiated by a local problem and the student's sensitivity to a social issue. They were presented at the end of two weeks of work over a few pdf slides. The presentations turned out uniquely just as the voices participating in this experiment. I took the task of post-producing the quick collage we created collectively to give it one formal language, but it could have as well remained six individual items if the students had more time to work on it.

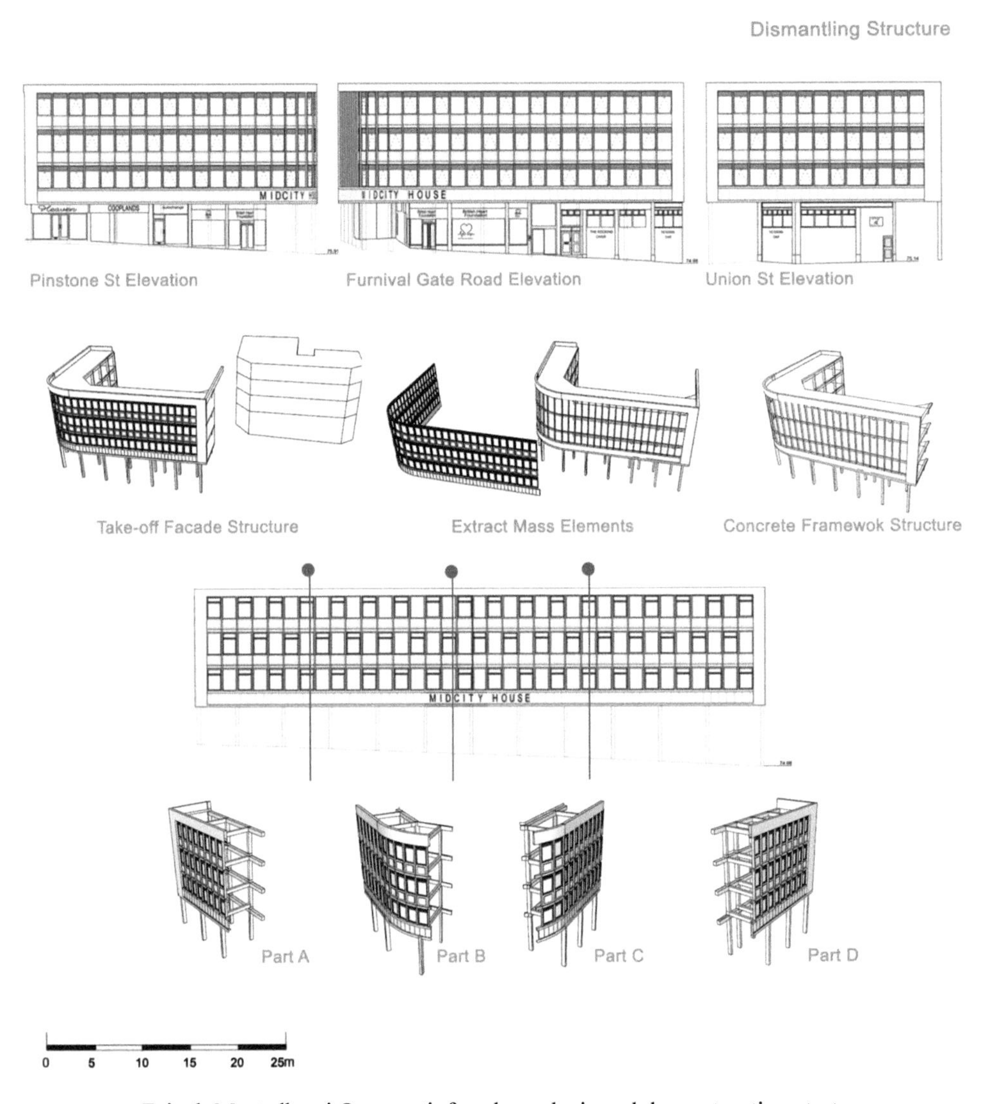

FIGURE 10.6 Zainab Murtadhawi Qanawati, facade analysis and deconstruction strategy

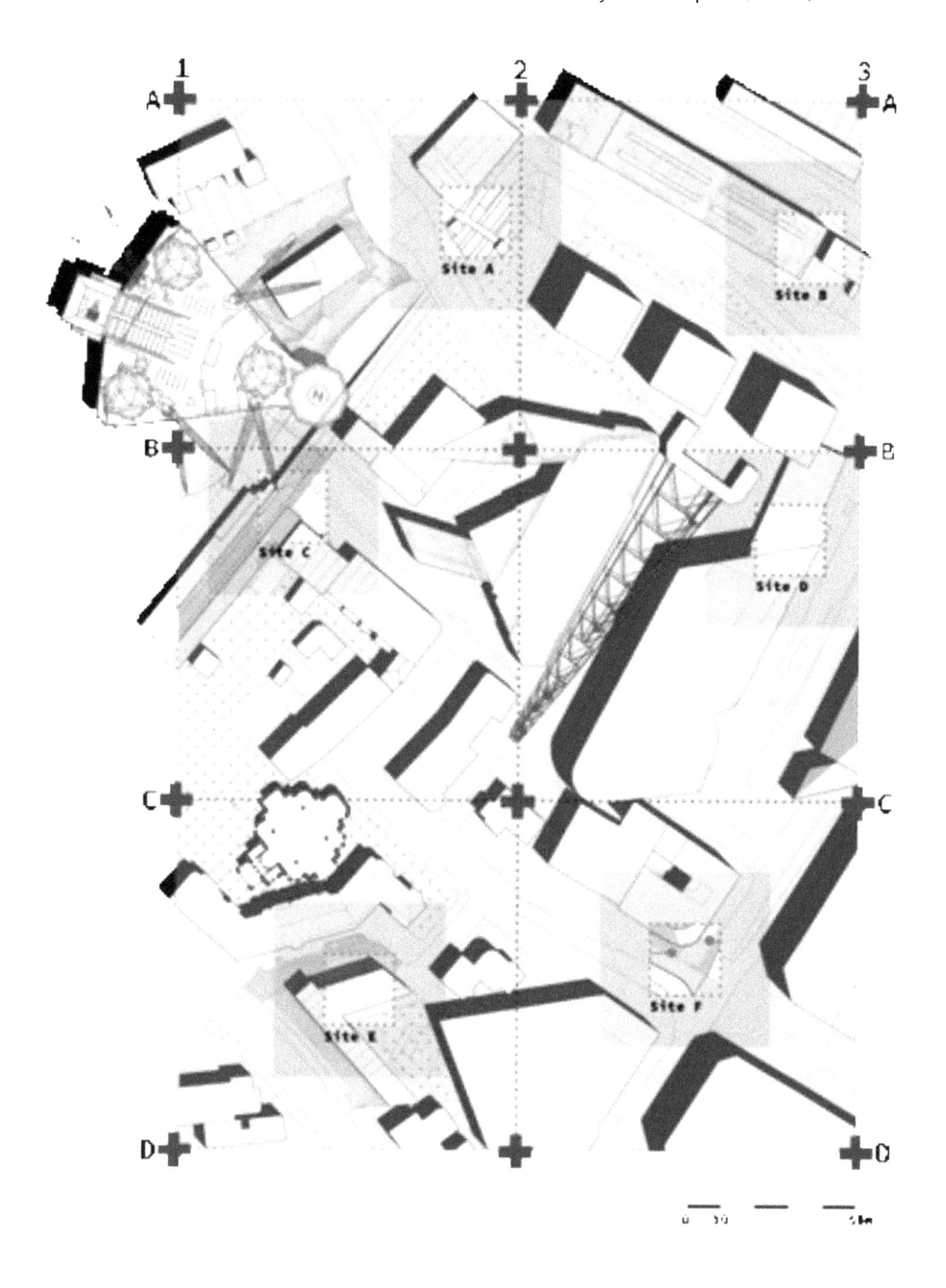

FIGURE 10.7 Team effort, Collage of six sites

Conclusions

In conclusion, this unique workshop, blending gamification and in-depth exploration of Collage City concepts succeeded in making the serious issue of urban transformation and sustainable material use more accessible and engaging. By challenging the traditional modernist notion of a 'tabula rasa' and embracing the rich diversity inherent in dense urban contexts, participants became both users and creators, contributing to a democratised approach to city-making. The need for democratising urban design was strongly conveyed and agreed upon. Furthermore, students became actively interested in demolition practices in their surroundings.

However, the workshop's hybrid format, combining online lectures and in situ explorations, presented its own set of challenges, from time constraints to limitations in fostering a sense of community. Despite these hurdles, the journey of discovery, intuition and collaborative learning prevailed, demonstrating that the process itself was a valuable and successful outcome, paving the way for future exploration of sustainable urban practices.

Sites

- Aidana Roberts: Farringdon, London, UK
- Ana Pastor: Becklow Gardens, London, UK
- Weronika Walasz: West Brompton, London, UK
- Zainab Murtadhawi Quanawati: Sheffield City Centre, Sheffield, UK
- Amila Strikovic: Strasbourg, Austria
- Anthony Roberts: Moulsecoomb, Brighton, UK
- Katarzyna Sołtysiak: Rotterdam, Netherlands

Notes

1 Oxford English Dictionary, s.v. "hybrid, n. & adj." April 2023. https://doi.org/10.1093/OED/8459251414
2 Rowe Colin and Fred Koetter, *Collage City* (Cambridge, MA: MIT Press, 1978).
3 Christoph Grafe and Jan de Vylder, *Bravoure Scarcity Beauty* (Antwerp: Flanders Architecture Institute, 2016).
4 Alastair Brotchie, *A Book of Surrealist Games: Including the Little Surrealist Dictionary* (Boston: Shambhala Publications, 1995).
5 Letter to the author, May 1991 in: Caragonne, Alexander, *The Texas Rangers: Notes from the Architectural Underground* (Cambridge, MA and London: MIT Press, 1993).
6 Colin Rowe and Fred Koetter, *Collage City* (Cambridge, MA and London: The MIT Press, 1978).

Bibliography

Brotchie, Alastair. *A Book of Surrealist Games: Including the Little Surrealist Dictionary*. Boston: Shambhala Publications, 1995.

Caragonne, Alexander. *The Texas Rangers: Notes from the Architectural Underground*. Cambridge, MA: MIT Press, 1995.

Colomina, Beatriz and Mark Wigley. *Are We Human?: The Archaeology of Design*. Zurich: Lars Müller Publishers, 2016.

Giddings, Joe. "Demolish Nothing: Densifying the Built Environment through Accretion." *Architectural Review*, August 2, 2023. www.architectural-review.com/essays/keynote/demolish-nothing-densifying-the-built-environment-through-accretion.

Grafe, Christoph and Jan de Vylder. *Bravoure Scarcity Beauty*. Antwerp: Flanders Architecture Institute, 2016.

Graves, Charles P. "The Plan Game: The Origins of 'Collage City'." *Looking@Cities*, September 6, 2018. https://lookingatcities.info/2018/09/05/the-plan-game-the-origins-of-collage-city/.
Imam, James. "Architects Dreaming of a Future with No Buildings." *The New York Times*, February 12, 2021. www.nytimes.com/2021/02/12/arts/design/superstudio-civa.html.
Jacobs, Jane. *The Death and Life of Great American Cities*. New York: Random House, 1961.
Oxford English Dictionary, s.v. "hybrid, n. & adj." April 2023. https://doi.org/10.1093/OED/8459251414
Rowe, Colin and Fred Koetter. *Collage City*. Cambridge, MA and London: The MIT Press, 1978.

11

OFFCUT 'A HEURISTIC OF FLOWS'

Michaël Ghyoot and Sophie Boone

Behind new goods

Most goods put on the market – including building materials – are presented in pristine condition. Extensive packaging techniques aim to give buyers the impression that they will be the first to enjoy the goods. The different layers of packaging, the absence of visible wear and tear, the homogeneous condition of the objects, the communication and marketing tricks – all these aspects serve this single purpose.1

Yet, before the object comes into the buyer's possession, it is likely to have passed through many hands. The more sophisticated the object and the higher the number of components, the more likely it is to have followed a long trajectory before ending up on the market.

It was to follow such trajectories that we wanted to invite the participants of our group of the School of Re-Construction (SoR-C).

Our group's thematic was 'Offcut'. Technically speaking, *offcut* refers to a very specific type of waste: an excess of material which is cut off a main element, most of the time to confer specific dimensions or properties to it. However, we wanted to slightly expand this definition and have a more global look at objects trajectories, looking at offcuts, of course, but also at other sorts of waste created along the way.

We believe this kind of investigation can be a valuable source of learning for architecture students (as well as for many other practices). Since their future profession is (still) mostly about allocating material resources to build things, diving into the entangled circuits of materials and the intricate rationales behind waste is a good starting point to rethink the role of the architect in the context of climate and environment crisis – which was an explicit objective of the SoR-C.

Drawing upon the particular context of this online summer school (pandemic obliges), we wanted to take advantage of the diverse locations of the participants and explore a multiplicity of local situations. Participants in the 'Offcut' workshop of the SoR-C were asked to start from artefacts that they collected in their everyday environment.

Through investigations, site visits, meetings with key actors and enquiries into literature, they were to unravel the trajectories of these objects: how were they produced, where do they originate from, how did they reach the place where they were found, what is their usual destination,

DOI: 10.4324/9781032665559-16
This chapter has been made available under a CC-BY-NC-ND 4.0 license.

and how and why have they been discarded as waste at some point in their trajectory? These forensic analyses aim at getting a better understanding of the practices and the narratives that affect our material environment.

This chapter mainly sets out the background of this approach while sharing some findings from the 'Offcut' workshop of the SoR-C.

Through the back door

The investigative approach we suggested is based on methods used by Rotor. Indeed, during its first years of existence (between 2005 and 2008 approximately), the association conducted a very large number of factory visits.2 What interested us then was not the finished products sold in gleaming showrooms but what came out of the back doors of production sites and ended up in containers, that is, industrial waste.

In the multifaceted world of waste, industrial waste is a special category. Its generation is closely linked to a production process that is largely predictable. This is in contrast to consumer waste, whose production is less predictable as it depends on users' decisions and the vagaries of its use. Industrial waste is also unique in that its volume is almost 30 times that of household waste.3

In this sense, offcuts – but also the many other types of industrial waste such as misfits, production failures, etc. – are a particularly valuable starting point. Offcuts are, as several participants in the group discovered, the inevitable counterpart of various production processes. For every hole drilled in an object, there are particles of material that are lost. For every cut, there is an offcut. In short, every object put on the market has its own ghostly double in the waste world.

FIGURE 11.1 Examples of industrial waste: steel swarf

Source: Rotor

FIGURE 11.2 Example of industrial waste: colour transition in the production of plastic crates.

Source: Rotor

The encounter with this type of waste raises many questions. They may arouse a utilitarian interest: what can we *do* with this waste? But these fragments also raise a deeper question: what can waste *tell* us about the organisation of the economy, the value systems that govern it and the worlds it creates – and destroys?

Careful fieldwork can help us understand the ramifications of the industrial complex and, therefore, imagine other circuits, other ways of creating value, other relationships to work and materials.

Tracing flows

In order to explore these flows and their possible reconfigurations, one needs to be adequately equipped. A few references can support such investigations.

The representation of the economy as a system of flows is not new. One can think of the work of the French physiocrat François Quesnay in his *Tableau économique*, first published in 1768. This diagram attempts to represent the way in which economic capital is reconstituted through exchanges between different sectors of activity (agriculture, industry, trade). For Quesnay, it is

Tableau Économique

Objets à considérer, 1°. Trois sortes de dépenses; 2°. leur source; 3°. leurs avances; 4°. leur distribution; 5°. leurs effets; 6°. leur reproduction; 7°. leurs rapports entr'elles; 8°. leurs rapports avec la population; 9°. avec l'Agriculture; 10°. avec l'industrie; 11°. avec le commerce; 12°. avec la masse des richesses d'une Nation.

DÉPENSES PRODUCTIVES relatives à l'Agriculture, &c.	DÉPENSES DU REVENU, l'Impôt prélevé, se partagent aux Dépenses productives et aux Dépenses stériles.	DÉPENSES STÉRILES relatives à l'industrie, &c.
Avances annuelles pour produire un revenu de 600^ll^ sont 600^ll^	Revenu annuel de	Avances annuelles pour les Ouvrages des Dépenses stériles, sont
600^ll^ produisent net	600^ll^	300^ll^
Productions		Ouvrages, &c.
300^ll^ reproduisent net	300^ll^	300^ll^
150 reproduisent net	150	150
75 reproduisent net	75	75
37.10 reproduisent net	37.10	37.10
18.15 reproduisent net	18.15	18.15
9...7...6 reproduisent net	9...7...6	9...7...6
4.13...9 reproduisent net	4.13...9	4.13...9
2...6.10 reproduisent net	2...6.10	2...6.10
1...3...5 reproduisent net	1...3...5	1...3...5
0.11...8 reproduisent net	0.11...8	0.11...8
0...5.10 reproduisent net	0...5.10	0...5.10
0...2.11 reproduisent net	0...2.11	0...2.11
0...1...5 reproduisent net	0...1...5	0...1...5
&c.		

moitié passe icy

REPRODUIT TOTAL...... 600 ll de revenu; de plus, les frais annuels de 600 ll et les intérêts des avances primitives du Laboureur, de 300 ll que la terre restitue. Ainsi la reproduction est de 1500 ll compris le revenu de 600 ll qui est la base du calcul, abstraction faite de l'impôt prélevé, et des avances qu'exige sa reproduction annuelle, &c. Voyez l'Explication à la page suivante.

FIGURE 11.3 François Quesnay's *Tableau Economique*, 1759 version

Source: Public Domain via wikipedia.org

the regular renewal of the 'stock' of agricultural products that really generates value – the rest of the economic activities consist mainly of ensuring their circulation.4

Towards the middle of the 19th century, we can think of the splendid representations of the engineer Charles-Joseph Minard.5

Like many scientists of his time, he was fond of statistics and eager to eloquently represent them.6 Therefore, he drew maps of the territory representing the circulation of the various flows of commodities (cattle, coal, wine, etc.), using arrows and maps to do this. The cartographic markers speak of the spatial movement of goods, while the thickness of the arrows represents the quantities of materials in circulation and their fluctuations over time and space.

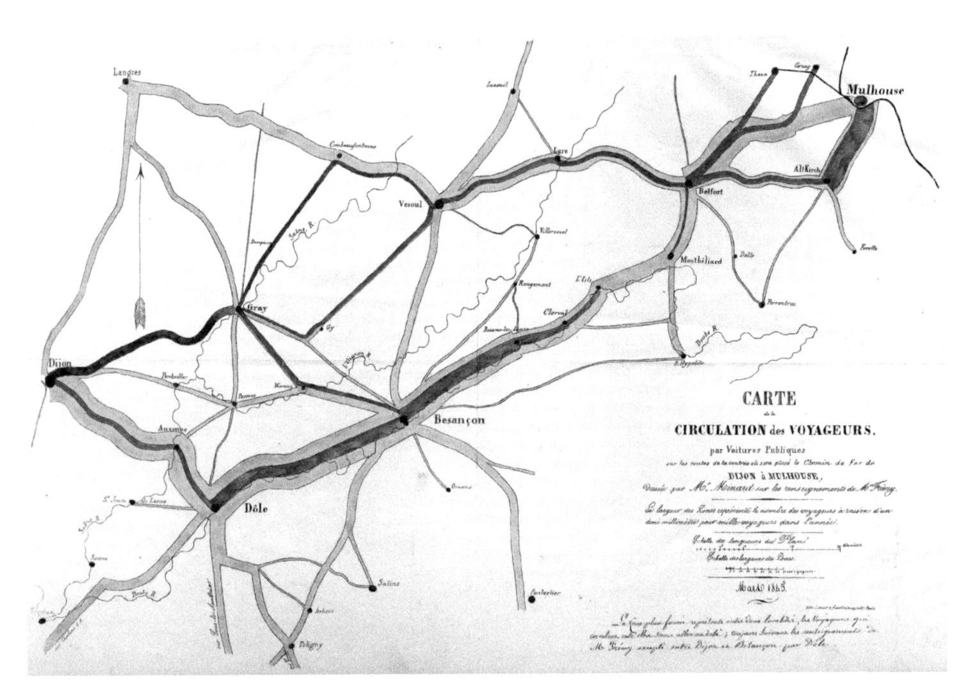

FIGURE 11.4 Charles-Joseph Minard's flow map of travellers by public coaches in Bourgogne, 1845

Source: Public Domain via wikipedia.org

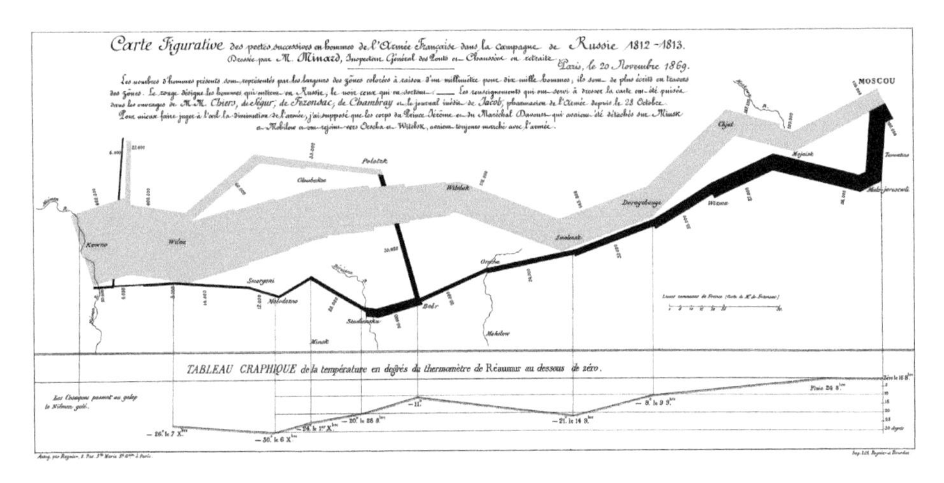

FIGURE 11.5 Charles-Joseph Minard's representation of the Russian campaign of 1812

Source: Public Domain via wikipedia.org

In another famous document, he uses the same method to depict, in a particularly striking image, the tragedy of the Russian campaign for Napoleon's armies. His diagram shows both the movement in time and space of Napoleon's troops but also their severe losses.

These "excellent space-time-story graphics illustrate. . . how multivariate complexity can be subtly integrated into graphical architecture."7

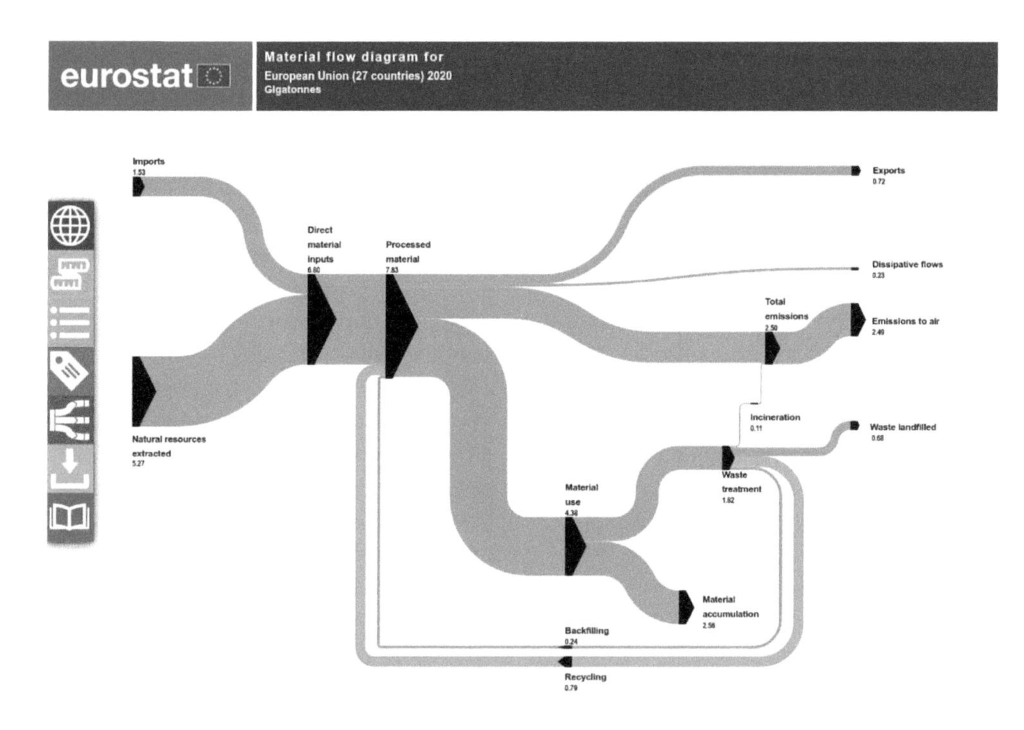

FIGURE 11.6 The material flow diagram produced for Eurostat, representing the material flows in European Union (27 countries) in 2020 (in Gigatonnes)

Source: European Union, Creative Commons CC-BY-4.0, via Eurostat

It was an Irish engineer who eventually gave his name to this type of representation: Sankey diagrams. Matthew Henry Phineas Riall Sankey sought to represent the circulation of energy flows within a steam engine and highlight the various forms of energy loss.

Sankey diagrams have become a classic format, widely used in scientific literature and in representations for the general public. Most recently, Eurostat, the European statistical institute, put online an interactive representation of material consumption in Europe based on a Sankey diagram.8

These representations were quite inspirational for the participants to the 'Offcut' workshop. They, too, attempted to represent the trajectory trough time and space of the items they collected.

Incidentally, with the work of Minard and Sankey, we already find several essential components of the current industrial economy: linear material flows and dependence on fossil fuel combustion.

From linear flows to circular economy

On the Eurostat representation, it is clear that the material flows consumed in Europe continue to be largely linear despite some modest feedback loops (mainly recycling activities). On a much smaller scale, the participants to the 'Offcut' workshop were able to make the same observation. Whatever the starting point chosen (a mask, a cigarette butt, a scrap of steel from a local metal workshop), the trajectories of the objects put forward have a largely linear appearance – not without interesting exceptions, as we shall see later.

In light of this, several schools of thought are calling for a transition to forms of circular economy.

In one of its strictest definitions, the circular economy postulates a profound paradigm shift. From a logic of flow production, we should now move to a form of stock management. Stocks are understood here as "assets of cultural, natural and human nature, and manufactured objects."9 According to Walter Stahel, one of the leading Figures in this movement of thought, "in industrialised countries . . . we have everything we need, but we have to learn how to look after it, and to care for it."10

The underlying idea is to close, or at least significantly slow down, the taps of mass production. By extension, this also means stopping – or at least significantly slowing down – the waste flows that are an inevitable counterpart of the production and consumption logics in a linear system. Stahel thus outlines an economic model in which prosperity ceases to derive from accelerated cycles of production-consumption-disposal but instead depends on activities of maintenance, upkeep, repair . . . in short, on preserving the use value of everything that is already among us.

On a theoretical level, this shift from flow management logic to stock management practices is not really new. Since the 1970s at least, thinkers and economists have been sketching out alternative models to the linear economy and the classical postulates on which it is based (ideals of infinite growth, availability and fungibility of resources, market equilibrium, etc.).

One of the strongest criticisms of the linear model comes from the work of the Romanian economist Nicholas Georgescu-Roegen, who was exiled to the US in the 1940s and whose most influential writings were published in the early 1970s. His reflections led him to strongly criticise the entropic nature of classical industrial economics:

> The economic process, . . . from a purely physical point of view, simply transforms valuable natural resources (low entropy) into waste (high entropy). . . . The fact that we are constantly drawing on natural resources is not without impact on history. It is even, in the long run, the most important element in the fate of humanity.11

Another example is Herman E. Daly, an American economist and environmentalist concerned about the massive consumption of natural resources. As early as 1974, he proposed a reversal of logic to get out of the contradiction between an ideal of infinite economic growth and a world with finite resources. He put forward a definition of efficiency understood as the maximum of services offered by the minimum consumption of raw materials:

> the services (satisfaction of needs) produced by the stocks of artefacts (and population) are the ultimate benefit of economic activity, and the consumption of raw materials is the ultimate cost.12

From this he deduced the following formula:

$$\textit{Final efficiency} = \frac{\textit{Service}}{\textit{Flows}} = \frac{\textit{Service}}{\textit{Stock}} \times \frac{\textit{Stock}}{\textit{Flows}}$$

According to him, this leads to two possible and complementary courses of action:

- Maintain the stock with less consumption of raw material (stock/flow ratio), which he calls maintenance efficiency. This is not infinite since, despite all possible efforts, the stock will inevitably deteriorate. However, it is possible to move towards a more efficient ratio.

- Produce more service from the existing stock (service to stock ratio), which he calls service efficiency.

These bold proposals by Daly, Georgescu-Roegen and others have not been translated into systemic change. Despite the impact of the Club of Rome report *The limits to growth*, the actual changes did not live up to the expectations. Debates on abandoning growth as a primary objective have been replaced by the notion of sustainable development.13 In a much more consensual way, it extends the *contradictio in terminis* of "sustainable growth."

Fieldwork enquiries to regain agency

The discrepancy between the circular and degrowth model and the current organisation of the economy is obvious. This is particularly the case in the construction sector. In Belgium alone, the production of demolition waste has doubled in less than 15 years.14 On a global scale, the consumption of natural resources has almost quadrupled since 1970.15 Similarly, greenhouse gas emissions have hardly stopped increasing for several decades. What is worse, emissions from human activity are accelerating.

The fieldwork enquiries such as those outlined by the 'Offcut' group's participants allow us to go beyond these Figures. They allow a better understanding of how these major trends are embodied in local and situated practices.16 The production and marketing chains are more complex than the big statistical Figures would suggest. By getting closer to the field and following material trajectories in detail, we can begin to discern situated practices that may, in some cases, suggest alternative trajectories to the paths of the current economy.

So, when a SoR-C participant based in Bath, UK, begins to investigate local stone (a fragment of which she picks up from a wall coping tile during a walk in the city), she uncovers the history, geography and preservation practices of this material. By following the trace of this small fragment, she ends up getting in touch with local companies specialising in the reclamation, conditioning and sale of these salvaged building materials. These actors, although in the minority today, are the perfect embodiment of the circular economy concept as defined before. They demonstrate that it can be prosperous to maintain existing resources in circulation. More broadly, she also discovers how former underground stone quarries in Bath are now being used for mushroom growing projects that contribute to a relocation of food production.

In the same vein, when another SoR-C participant asks what happens to the wooden planks that make up the scaffolding in front of her Brighton home's window, she discovers, on the one hand, the globalised management of wood waste flows (crushed to make particle board in a Gatwick factory or incinerated in furnaces in Sweden) and, on the other, the hyper-local, largely community-based practices of upcycling and re-use.

Or, when another Brighton student, also taking part in the 'Offcut' workshop, collects a used pillow from a skip at the foot of his student residence, he holds the first clue in an investigation that will lead him through the globalised production circuits of low-cost consumer goods by Ikea, through the (no less globalised) circuits of waste management by Veolia or through the maintenance practices of bedding elements in the hotel sector. In the course of his investigation, he will also be confronted with the subtleties of the composition of pillows and its influence on the (new) value of these elements and the organisation of re-use pathways. Finally, refocussing on his starting point, he will initiate discussions with the university and the organisation

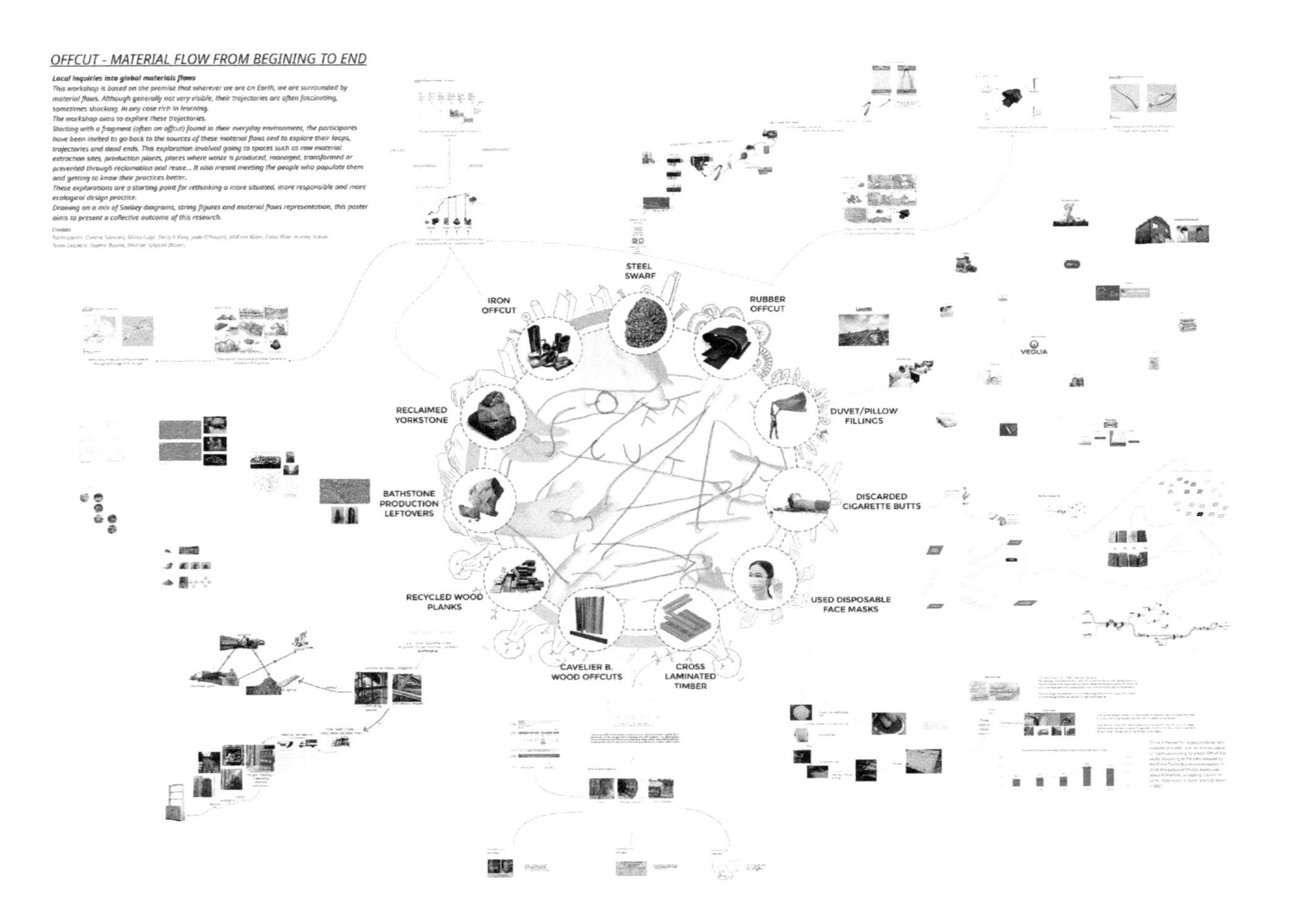

FIGURE 11.7 Final presentation of the results of the investigations carried out by the participants to the 'Offcut' workshop of the SoR-C

Source: School of Re-Construction, group 'Offcut'

in charge of managing the built infrastructure to develop a re-use scheme for bedding elements when students leave their residences.

The same approach was followed by all the participants to the 'Offcut' workshop, sometimes for rather mundane objects such as surgical masks and cigarette butts. Other times, it was about production waste generated in workshops and factories (rubber scraps, steel swarf, steel and wood offcuts). Although necessarily limited by time and the practical constraints of working from a distance, as some participants were in a lockdown regime, all these enquiries have, nevertheless, enabled a better understanding of the intricacies of economic circuits. Most of them also identified possible branching points: small forks leading to other practices more respectful of materials, people and the environment.

Towards an ethic of 'response ability'

Work by Georgescu-Roegen, Daly, Stahel and others prompts dreams of a transition in the construction sector. Rather than being a nodal point in the consumption of raw materials, the production of waste and other environmental damage, the sector would bring together a set of practices focussed on maintaining, servicing and repairing existing buildings. Materials that are no longer needed would be reclaimed for re-use, therefore minimising waste. Architects would apply all their creative intelligence to adapting existing buildings to new needs while simultaneously reducing the material impact of their interventions.

Although we are still a long way from this, the work carried out by the participants in the SoR-C is, nevertheless, pointing in promising directions, likely to give substance to these ambitions.

Finally, taking the time to explore the complex flows of the material economy also means paying attention to the beings – both living and nonliving – that ensure their circulation. This increased attention is fundamental to an ethic of care, which inevitably accompanies a 'doing with' state of mind.

More broadly, one can dream of a connectionist and ecologist ethic of architecture in which design and spatial intervention decisions (including the decision to *not* intervene) would be based on a responsibility to answer to the multiple stakeholders involved in the supply, implementation and use of materials and spaces. Against the Roarkian17 Figure of the architect alone against all (who ends up dynamiting his own building under the pretext that it has been corrupted by concessions granted to his clients), it is a question of repopulating the design and construction sites with all the actors concerned, closely or remotely, by these questions. It is also by revealing these connections and close interdependent relationships that we will be able to imagine new economic forms and a fundamental renewal of our relationship to the built environment.

Notes

1 Rotor (T. Boniver, L. Devlieger, M. Gielen, M. Ghyoot, B. Lasserre, M. Tamm), A. d'Hoop, B. Zitouni, Usus/usures. *État des lieux – How things stand* (Brussels: Éditions de la Communauté française Wallonie-Bruxelles, 2010). Available online: https://rotordb.org/en/projects/usususures-etat-des-lieux-how-things-stand (last consulted on 24/10/2022).

2 Some findings of this research can be found in T. Boniver, L. Devlieger, M. Gielen, M. Ghyoot, B. Zitouni, *Deutschland im Herbst. Exhibition Catalogue* (Kraichtal: Ursula Blickle Stiftung, 2008). Available online: https://rotordb.org/en/projects/deutschland-im-herbst-0 (last consulted on 24/10/2022).

3 Although accurate data is hard to collect because of unmonitored flows. See M. Liboiron, "Municipal versus Industrial Waste: Questioning the 3–97 Ratio", *Discard Studies*, 03/02/2016, https://discard-studies.com/2016/03/02/municipal-versus-industrial-waste-a-3-97-ratio-or-something-else-entirely/ (last consulted on 24/10/2022).

4

> [The physiocrats] felt, in a vague way, that human labour is the source of all improvement in the standard of living; but on closer examination they found that human labour does not increase the number of things that exist, it merely changes their form: they concluded that labour does not really create new things. The Earth, on the other hand (and more generally Nature), creates new things, it is the source of a real 'over-product'. One kilogram of seed turns into twenty kilograms of harvest; twelve cattle turn into twenty or more after a year.
>
> *F. Vergara, Les fondements philosophiques du libéralisme. Libéralisme et éthique (Paris: La découverte, 2002 (1992)), 147 (author's translation)*

5 S. Rendgen, *Le Système Minard. Anthologie des représentations statistiques de Charles-Joseph Minard issues de la collection de l'École nationale des ponts et chaussées* (Paris: Éditions B42, 2020).

6 A. Desrosières, *La politique des grands nombres. Histoire de la raison statistique* (Paris: La Découverte, 2010).

7 E. R. Tufte, *The Visual Display of Quantitative Information*, 2nd edition (Cheshire, CT: Graphic Press, 2001), 40.

8 Eurostat, "Material Flow Diagram", 2021. Available online: https://ec.europa.eu/eurostat/cache/sankey/circular_economy/sankey.html (last consulted on 24/10/2022).

9 L. Gravis (ed.), "Walter Stahel and Ellen MacArthur – in Conversation", *Medium*, 26/07/2019.

10 Ibid.

11 N. Georgescu-Roegen, *La décroissance. Entropie, Écologie, Économie* (Paris: Éditions Sang de la Terre, 2011), 72–73 (author's translation).

12 H. E. Daly. "L'économie de l'état stationnaire" (1974), translated and quoted in D. Bourg, A. Fragnière, *La pensée écologique. Une anthologie* (Paris: Presses Universitaires de France, 2014), 396 (author's translation).

13 L. Devlieger, L. Cahn, M. Gielen, *Behind the Green Door. A Critical Look at Sustainable Architecture through 600 Objects* (Oslo: Oslo Architecture Triennale, 2014).

14 From 11 million tonnes in 2004, it went to 22,5 million tonnes in 2018. Eurostat, "Generation of waste by waste category, hazardousness and NACE Rev. 2 activity."

15 W. U. Vienna, *Material Flows by Material Group, 1970–2019. Visualisation Based Upon the UN IRP Global Material Flows Database*, (Vienna: University of Economics and Business, 2022). Available online: www.materialflows.net/visualisation-centre (last consulted on 20/07/2022).

16 A. L. Tsing, *Friction: An Ethnography of Global Connection* (Princeton, N-J, Woodstock, Oxfordshire: Princeton University Press, 2005).

17 Howard Roark is the hero of Ayn Rand's novel *The Fountainhead* (1943). He truly embodies the modern architect ethos.

12

MATERIAL FLOWS

Nicole Maurer and Mark Oldengarm

FIGURE 12.1 Graphic representation of the core elements of 'material flows', made by the students

FIGURE 12.2 The students in stick-Figures, made by themselves

DOI: 10.4324/9781032665559-17
This chapter has been made available under a CC-BY-NC-ND 4.0 license.

Introduction

As occasional partners, we did not have a specific method or way of working we could put into practice. What we did have was our conviction that we could guide our students through the complexity of 'flows' within a given situation and be able to have them make use of their own talents and interests to investigate the broad meaning and implications of re-construction.

We started thinking how we could formulate that into an application for the summer program School of Re-Construction. Quickly, we agreed to not primarily focus on the physical representation of what you could call a 'circular architecture'.[1] Instead, we wanted our students to look beyond buildings and/or their materials, for they are the product of many intertwined processes that very often have historical, cultural or vernacular connotations. By taking into account all kinds of aspects on the 'meaning' (relations) in circularity, they might find out that sometimes that can be more important than the actual 'function' (goal).

A re-constructed architecture may be different from what people or politics are used to encounter as the result of a process. The different aesthetics will affect the acceptance and should, therefore, be an issue in the process of re-construction. So, re-constructing, in our opinion, means also to de-construct on a symbolical level, retrieving meaning, history and culture. These elements will support the acceptance because of the 'value' they add in being recognisable and able to relate to.

We wanted our students to practice in new ways of thinking, connecting products and ideas into a plausible, coherent and convincing narrative. If a circular story that defines identity becomes embraced by public or politics, you have actually created a sustainable asset.

Preparation and brief

For our preparations, we sat at a terrace on the banks of the Meuse in Nicole's hometown, Maastricht, Netherlands. It was in the week after large floodings had their severe impact in several European countries. The Meuse was still at a very high level and contained mud and rubbish from the damage done upstream. This situation made the subject of climate change tangible and the question of human adaptation to it felt urgent. How can we possibly construct our buildings in the vicinity of such natural powers, knowing that the circumstances of heavy rainfall and rising water levels are an absolute threat to our state of living?

We needed other perspectives, other stories, different kinds of thinking to help us approach major questions like this. 'Take the old quarry, for example,' said Nicole, 'it is meant to be redeveloped, because it is situated on a beautiful location. But also a very vulnerable one.' It did not take us very long from that realisation to formulate our basic concept for the School of Re-Construction.

Coincidentally, our respective hometowns were in the process of redeveloping an area that had facilitated the growth and welfare of broad surroundings in the second half of the 20th century – in Zwolle, Netherlands, the former gas-powered energy station IJsselcentrale (also known as the Harculo-Centrale) and in Maastricht, it is the quarry of Eerste Nederlandse Cement Industrie/First Dutch Cement Industry (ENCI). What made them suitable is that they have several elements in common but are in different phases in time towards a future with new functions to be developed.

The common grounds are that both locations can be found just outside the dense urban area of a mid-size city (to Dutch understandings) and are situated directly next to a river. Both had

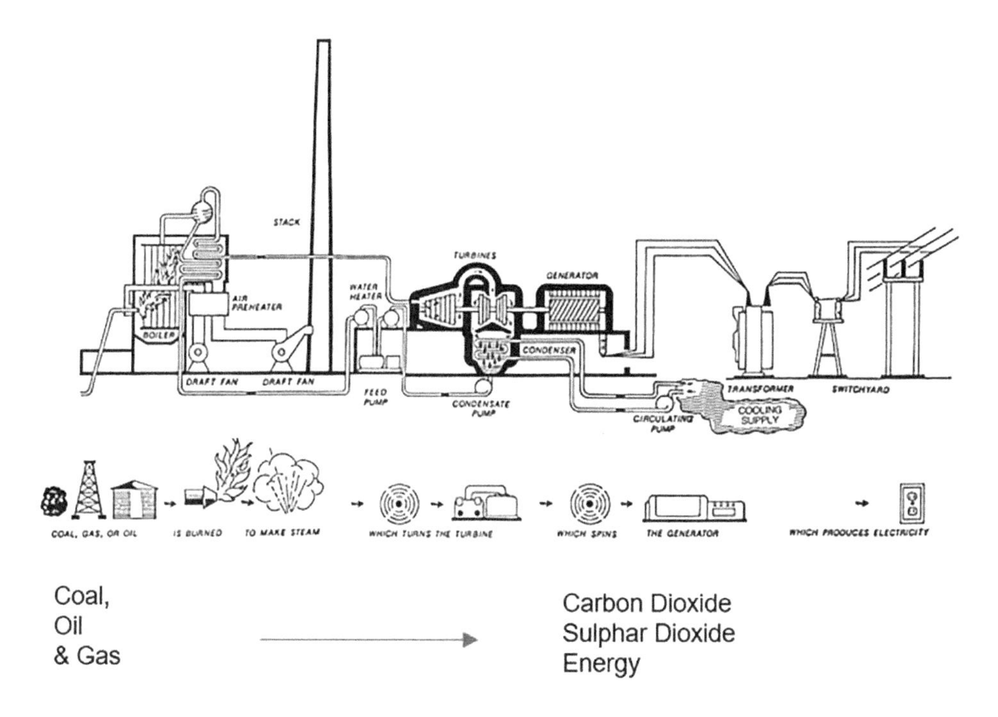

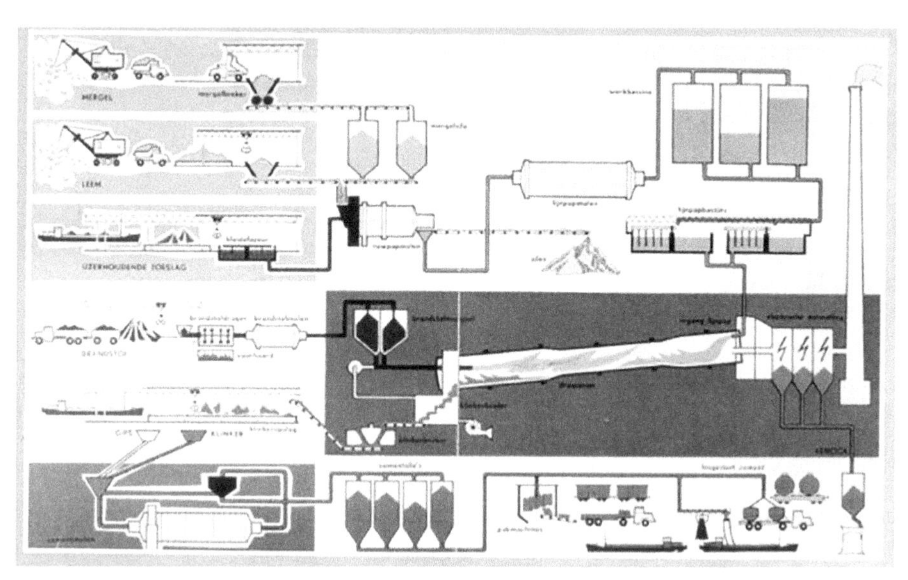

1 Overzicht van de verschillende stappen in het productieproces (natte procedé)

Limestone / Marl → Carbon Dioxide
Cement

FIGURE 12.3 These Figures show the processes of producing electricity (a) as the IJsselcentrale did and producing cement using the so called 'wet process' as was done by ENCI (b) a function as production areas that facilitated (the growth of) the Dutch built environment and welfare in the 20th century. The raw material for building houses came from the ENCI quarry, where the IJsselcentrale generated electricity that made houses very comfortable to live in (see Figure 12.3). There is a clear opposition between the former use of fossil materials (gas, cement) to improve welfare and the contemporary lines of thinking that aim more at the general improvement of well-being for man and nature. That brings us to the last element both locations have in common; they are surrounded by important natural areas (partly Natura 2000, which is the highest category of protected nature).

The theme for our group was 'Material Flows', a nice combination of the physical (materials) and ephemeral (flows). In order to have a bit more grip on what material flows could mean, we combined a couple of definitions that we found in literature or online that fitted the notions we came up with in our introduction:

> The 'material flows' are generally defined into water, energy, materials (including food), and waste. These flows consist of inputs (local, regional, and global inflow of resources), throughputs (energy required to transform these resources and waste produced by any process), and outputs (the material outcome of this process). [2]

This seems like a technical quote, but if you leave out the notion of the resources being physical, then the 'flows' can be cultural, human, natural or time related as well and offer a framework for stories. People can relate to stories, recognize themselves or their situation in it and, possibly, come to a better understanding of the complexity of things. This is a process that will normally take quite a long time to reach some kind of destination. Therefore, we kept in mind a quote that was probably from Stewart Brand and relates to the aspect of time and impact: "*Fast gets all our attention, but slow has all the power*".[3]

A short brief to our students

For our students, we wanted to make sure of two things; the first was that we wanted them to experience the value of spending lots of time on investigation so there was no specific or goal-oriented thinking but, instead, quite an open assignment to enable them to make room for possible unknown perspectives and connections. The other thing was that we wanted them all to do was bring something personal into the project. That could either be something they thought of upfront for themselves to learn or develop, it could be a new subject they felt connected to within the first couple of days or it could be a deeper dive into elements that already had their interest.

During the process, we specifically asked them to make a connection to any existing experience they have. For they all are inhabitants, citizens and humans with a certain cultural background. And for this matter, they had very different backgrounds coming from all over the world. That does matter and also directs to a certain extend their approach. The fact that they also are trained specialists is very useful luggage and should, of course, be put into practice, but we were looking for what their stories could be.

Prior to our first meeting, we briefly shared our thoughts on re-construction and storytelling in an email and explained the value we see in the sense of place as a sustainable asset. The only particular result we named is that the outcome of the two weeks should not be a solution to a problem or an answer to a specific question. What we wanted them to achieve was processing their talents and knowledge into a story that is convincing enough to the public and/or politics. They should be triggered to transfer their mindset into one where circularity, sustainability and nature-inclusive design engagement are logical and necessary elements in undertaking an area development or gaining societal well-being.

In short, it should show an attractive and possible sustainable future. The aim was to define a visual storyline, a scenario of challenges, concepts vision and ideas. This outcome should be presented as a new flow in the form of some sort of digital 'exhibition' at the end of our two-week course. They put it like this themselves:

> We were asked individually and collectively to study the material flows of *two sites*, in *two cities* within the Netherlands. Our challenge is to improve site by observing the *many material + immaterial flows*, using narrative as a tool for producing a convincing proposal for People, Partners & Policy Makers (& Planet).[4]

Guiding process

This quite short brief was all our students received before they were asked by us to write down their own thoughts in a short motivation in preparation to our starting session. That gave us two things; one was the small start of stories, for everyone got to know a bit about each other's background. The second thing was that they almost all came up with general and broadly orientated input, but they also posed lots of questions like "Is it also possible that I . . ." or "Is the subject of . . . possible to fit in?" To us, this meant we had the open and investigative minds we hoped for, and we found personal connections to the theme.

Following the short brief, we wanted to make our students benefit as much as possible from the curiosity and inquiring minds they already possess. So, we provided only a compact overall scheme for the two weeks, which divided their process roughly in two components. The first week was about what we called 'collect, combine and define' and the second one about 'design and share'. The students started doing desk-research by asking themselves what it is they needed to know and, in the following phase, they exchanged and combined the diversity of information to define a direction for themselves. That direction did need to be incorporated in the group story, of course, so they needed to discuss their findings and ideas with the group.

To facilitate the process of investigating and deliberating, there was a daily meeting with the group leaders, meant to discuss outcomes and progress and next to that we made individual appointments to make sure that everyone kept their personal connection to what they were doing. These individual conversations also proved to be useful considering the fact that there were differences in experience and knowledge between the students. It made us able to have attention for the individual progress while at the same time making sure that the contribution to the group process was secured.

Content development

At the start, we presented the students general information on the locations and a couple of basic outlines to work with. During the week, we provided them with additional theories or information that related best to the phase they were in at that time. On the one hand, these were methods that had to do with sustainability and circularity (e.g., the building layers of Stewart Brand,[5] the

10-R ladder,[6] the 'First Guide to Nature Inclusive Design'[7] or a piece on how to transform cities into thriving circular societies) and, on the other hand, it was basic working theories (e.g., visual thinking's methodology,[8] the technique of Futuring,[9] the Theory of Change).[10] The next days the students got to dive into the material and write down their interests, findings and questions on the common canvas (Miro) to be discussed in our daily meetings. During the week and in our meetings, they got acquainted with each other better and frequently worked in smaller teams as well.

A lot of aspects of both locations and of general sustainable themes were taken into account, turned around, de-constructed and finally spread out in small elements all over the common canvas. Then the combining started and groups of information or thoughts started to come together in the minds of our students. A few examples – by putting into practice the top step of the 10-R ladder (refuse), a student proposed to hand back the locations to nature. By doing very little (footpaths, viewing point) natural qualities are added. Another is that by using an existing connection, the concept arose to create an iconic renewable energy plant. In renewing the site heritage from a sustainable vantage point, the idea was to create a possible engine for the transformation of energy supplies. A third one came up with the development of a general matrix for area development to create insights in which directions can be successful to investigate more in depth.

Next to these examples, there were a lot of other ideas. Many of them contained re-use of materials that were on hand at both locations. Secondly, water was present for use generating renewable energy and as a recreational factor. And thirdly, the value of the well-being of people and nature played a large part in the students' search for encountering the sites. They were all looking for elements of re-generation or reviving. Their common goal became to develop thoughts that not only improve quality of life (both human and nature) but also offer perspective for the future.

This actually was not solely driven by the students' ideals. They found elements of their own idealistic views in the origins of both locations. As said before, the locations contributed to the welfare of the Dutch in the second half of the 20th century. The Harculo power plant was made out of a very innovative concrete (shock-concrete), which was very strong and long-lasting. The construction was largely done in repetitive panels which made building quick and easy. It was at the core of the many farms in the newly formed province of Flevoland (fully reclaimed from an inner sea). One student found out that ENCI actually became the major shareholder of the shock concrete company. That triggered new storylines, as connections were seen everywhere.

It was the combination of water, material flows and the idea of forming a better world that led to the first main carrier of the story. A couple of students checked out the infrastructure of rivers and found out that there was a possible route to travel between Zwolle and Maastricht. One student even started looking into the income rate of people living alongside the route when the idea arose that a vessel could bring stuff and food between the locations. We really loved that one. And from the idea of the vessel, it was a small step to give a twist on the biblical idea of the ark of Noah. As a recognisable Figurehead, quickly the idea arose to use the very Dutch icon of Nijntje as a symbol of these twin cities. Nijntje – Miffy in English – helped children to relate to certain situations and is a worldwide known character from the children's book series by Dutch artist Dick Bruna.

To elaborate on the ark of Noah our students came up with a boat. Powered – of course – by renewable energy and with sustainable ideas, knowledge, healthy food or even plants and seeds (inspired by the Seeds of Change project)[11] as cargo for every village alongside its route. Both

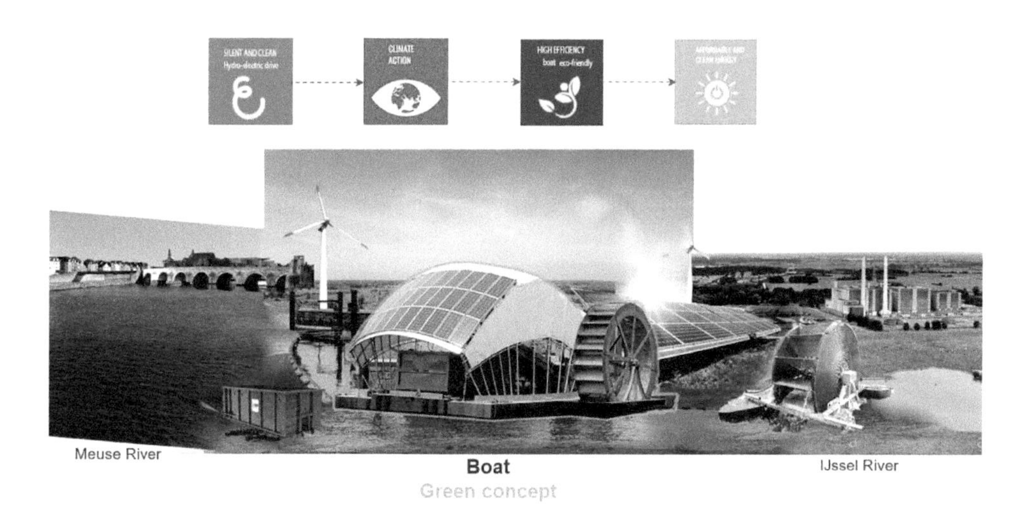

FIGURE 12.4 Representation of the green concept 'Boat', inspired by the seeds of change project

Source: Illustration made by the students

the locations in Zwolle and Maastricht in this storyline function as (iconic) beacons or frontrunners for possibilities to create a sustainable, social and nature-inclusive way of living.

One student proposed to design a Monument of Change as a physical representation of the beacon idea. The monument could contain new ways of agriculture and the re-use of materials from either site and of nearby industries. Or room for bird watching and nesting, a collecting instrument for rain water and, of course, the vessel could moor at its base. The student explained that, to her, the icon had elements in it of how the Dutch are used to exploiting every inch of land they have in a way that it has the highest economically efficient output while, at the same time, contributing to awareness and the narrative.

Stabilizing the storyline

So, the first week was all about collecting. The second week proved a bit more difficult. The individual processes and the group project were not always easily to be aligned. The students sometimes struggled to keep up the idea of doing justice to both as did we. Where the online part worked really well in Week 1, it had its negatives as well. The 'design and share' phase is about working out the different ideas and keeping an overall clear storyline as well, which requires lots of group communication. But more than one person speaking in an online environment is confusing and different time zones were an obstacle as well.

Nevertheless, we think the outcome did right to our students and the group result. This had to do with the process of Week 1 where exchanging bits of personal interests and background took place. Another element was the fact that the summer school did not require two weeks of full-time participation. The students fitted the schedule into their regular business. So, during our working period as a group, we encountered each other's specific interests, jobs, backgrounds and more. Also the ideas that came to mind found their way back into the daily business of our students. It brought about questions and perspectives that we think would not appear in the same amount when collaborating in a workshop setting. It was somehow daily life entering our process.

Here again, it proved to the students the importance of stories. Or 'daily flows' as someone called it. They could relate it very well to the process of indexing existing values as it was done in the Superlocal project[12] by Maurer United Architects. We encouraged sharing what was on everyone's mind or what had set their brains at work. The practical issues were difficult, but this situation brought its own values as well. The approach changed more than once from architecture orientated to questions like food security, housing, poverty, rising water levels and biodiversity. This still left us with trying to converge the process that was diverging into these bigger (global) urgencies.

Maybe some sort of scheme would help? The idea of using a simple methodology that one of the students developed did not resonate at first, but when everyone actually filled in the scheme, they were proposed with easy-to-read diagrams on their own approaches and values, they attributed to both sites. It proved helpful indeed by determining why everyone was so involved with what they came up with – not a solution, but very helpful insights.

When viewed from the perspective of the larger issues, it turned out that the two locations we offered our students were much more than just interesting areas that could be re-developed. Collecting and combining information through the lens of these major questions gradually formed a promising storyline: *A Tale of Two Cities, Two Sites and Many, Many Flows*

> *It was the best of times, it was the worst of times, it was the age of wisdom, it was the age of foolishness, it was the epoch of belief, it was the epoch of incredulity . . .*[13]

As said before, we had intelligent and inquisitive students. We also had a group that diverted in the way they preferred to put their idealism into practice. Yet, they managed to create a coherent story with enough space for everyone to elaborate on their specific input or process. The fact they all delivered a part of the final presentation was proof of their teamwork.

The carriers of their story were Miffy and a vessel that interlinked our two sites and was aimed at the regeneration and sharing of nature and knowledge. The ladder of the 10-Rs as an

Analytic hierarchy process

A structured technique for organizing and analyzing complex decisions, based on mathematics and psychology.

Pairwise comparison is a process of comparing alternatives in pairs to judge which entity is preferred over others or has a greater quantitative property. Pairwise comparison is one of the ways to determine how to access alternatives by providing an easy way to rate and rank decision-making. It is often used rank criteria in concept evaluation.

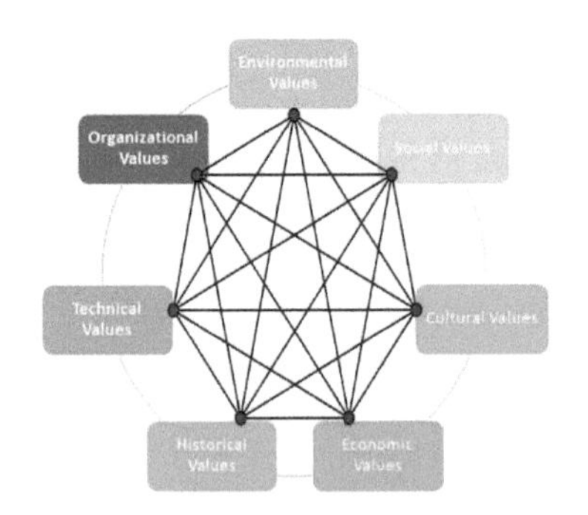

Ranking and Weighting the Criteria

T. L. Saaty, "The Analytic Hierarchy Process: Decision Making in Complex Environments," in *Quantitative Assessment in Arms Control*, Springer US, 1984, pp. 285–308.

FIGURE 12.5 Scheme of the analytic hierarchy process, which the students used to clarify and compare their individually scored values in relation to the project underlying structure helped in keeping it coherent; however, it did not stop them from using other instruments (like the Sustainable Development Goals) in corroborating their proposal towards what they named 'Re-constructing, Re-wilding and Re-thinking'.

Starting by making use of Miffy to explain our current handling towards our planet, they quickly moved towards some of the tracks they explored: from re-use to restore and letting the flow of time be the architect of a new landscape. This formed the basis of the three different minimal approaches (as opposed to a maximum one): Landscape, Tower and Boat (Figure 12.7). The Landscape approach considered each site as a chance for refusal and reclamation, where the Tower approach was aimed at re-visioning climate change and re-vitalising ecology. Thirdly, the Boat was a practical and metaphorical vessel for re-thinking relationships on each site as a carrier of materials, knowledge and narrative.

FIGURE 12.6 Miffy visualizing the way we handle our planet by using a jackhammer to mine into a (circular) flow

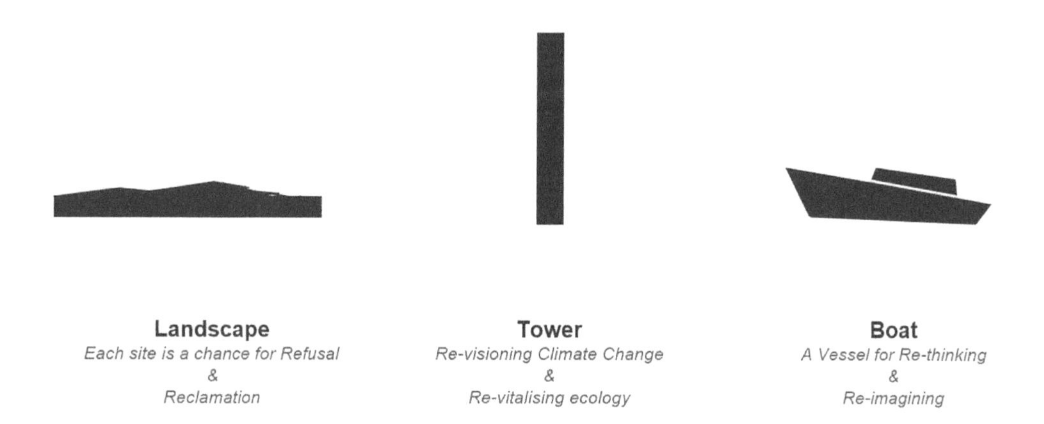

FIGURE 12.7 Three minimalized Figures to depict the students' approach on material flows

Each one was elaborated on through one or more projects of our students. Landscape made one of them say that "Refusal is a way forward to a new fresh landscape "where" we will witness a new flow that will surprise us in unexpected ways: seeing the materials live, proper, decay and die". Somehow this reminds of Romantic ideas and the concept of opening the sites up for leisure combined with ecological re-connection as a *natural public resource*.

We've already described the Tower (Monument of Change) and the Boat (A vessel for re-thinking and re-imagining), but we need to end with the fantastic and thorough work of Week 1. From the usual material flows our students dug into, they came up with ideas that indeed could spark people, partners, policymakers and the planet.

Conclusions

"How to teach architecture in an age of climate emergency?" That was an underlying question during the School of Re-Construction. And the answer is probably too complex to offer a blueprint sort of solution. What we have tried in our group is not teaching architecture itself but creating a process that was as open as possible in addressing the 'bigger issues'. So it was more about guiding than teaching, more about addressing than solving and more about re-search than re-design.

We wanted our group to experience that the playing field of architecture does not merely consist of different designing questions, rather it is a very complex collection of interests that both overlap and contradict.

Invest and investigate

The approach to such complexity is not an easy one. As for the process, we can sum it up in two words: invest and investigate. Coming up with possible designing answers or storylines for complex questions takes a lot of time. The majority of work needed for a circular process is preparation, such as making sure you know what you are talking about that all stakeholders and interests have been taken into account. The amount of time needed to carefully follow through such a process brings us back to the earlier mentioned quote that "slow has all the power". We can only hope that stories like the ones our students came up with are, indeed, powerful enough to spark some people to take an interest or possibly even come into action with regard to sustainability.

But we mainly hope that our students will take with them an approach they can put into practice somehow in their daily business of being an architect(-ure student). For we think that architecture can play a more important role by visualising storylines in dealing with urgencies of today and tomorrow. This means that a large part of the designing and analytic capacities should be invested in investigating the underlying 'flows' that are present in almost any project.

Notes

1 By circular architecture, we mean an architecture that is able to be repaired, re-used or regrown when needed. No harmful raw materials will be used or it will re-use already produced elements in its design.
2 This is our own combination of several definitions from different sources.
3 The essence of this quote stems from *The Clock of the Long Now* by Stewart Brand. A short explanation can be found at this webpage: www.conversationagent.com/2020/09/fast-gets-all-our-attention-slow-has-all-the-power.html

4 The words in italic were done so by the students to mark the essence of what they where digging into.
5 See, for example, this article by Stewart Brand: https://jods.mitpress.mit.edu/pub/issue3-brand/release/2. It explains briefly the notion of pace layering, which he coined in his 1994 book *How Buildings Learn: What Happens After They're Built*.
6 The 10-R circularity framework was co-developed by Jacqueline Cramer, former Minister of Housing, Spatial Planning and the Environment. Some background can be found on page 15 of this document (in Dutch): https://platformcb23.nl/images/downloads/20190704_PlatformCB23_Framework_Circulair_Bouwen_Versie_1.0.pdf
7 The guide can be downloaded at: www.dsla.nl/en/nieuws/the-first-guide-for-nature-inclusive-design/
8 See, for example, this website: https://visme.co/blog/design-thinking/
9 See, for example, this piece: www.tandfonline.com/doi/full/10.1080/03098265.2020.1869923
The book *Wijzer in de toekomst: Werken met toekomstscenario's* (in Dutch) by Jan Nekkers is very informative as well and provides a step-by-step guide.
10 www.theoryofchange.org/what-is-theory-of-change/
11 The project 'Seeds of Change' is about origin and transportation of seeds, visualized in a floating garden. A brief explanation and pictures can be found at: www.dezeen.com/2012/09/14/ballast-seed-garden-by-gitta-gschwendtner-and-maria-thereza-alves/#
12 The Expo-pavillion of the Superlocal project was done by Maurer United Architects. It also conducted a social investigation amongst local inhabitants on how they view their living environment. The findings of these projects were also lectured by Marc Maurer at the School of Re-Construction. Several publications and more backgrounds can be found at: www.superlocal.eu/
13 Opening fragment from the first sentence of *A Tale of Two Cities* by Charles Dickens. The title of the final presentation was inspired by it.

13

HOUSING LIFE CYCLE EXTENDED

How to identify, adapt, and re-use existing buildings and their components to support housing

Taleen Josefsson and Filipa Oliveira

Introduction

In most recent memory, architects and designers are both taught in their foundational architectural training to design projects where often materials appear to be disembodied from their origins, stories, and properties to fit into heavily theorised forms and the clean lines of digital design programs. This approach reflects neither the realities of the heavy extractive material resource demand on which this type of disembodied design relies nor the negative environmental impact that results all too often in the increase of greenhouse gas emissions, watercourse pollution, air and noise pollution, waste generation, and material consumption (Chuai et al., 2021).

Currently, housing design is repeatedly planned to respond to strict regulatory frameworks and to fill shortage requirements which have led to the delivery of a housing typology unresponsive to local conditions, unadaptable and constructed to suit a specific time (Beadle et al., 2008). These typologies are ultimately unsuited to the local climate which is becoming more and more global. More often than not, imported high environmental impact materials are prioritised over locally available, less environmentally damaging resources grounded in the geography and history of a place. This is a systemic issue, and while blame does not solely rest on architects (nor any singular party), these flawed practices have produced a global problem, where the lack of high-quality, long-term affordable housing calls for a new approach and reform of statutory legislation.

If adapting and re-using existing housing structures is encouraged in design briefs and given preference over new de-contextualized buildings, we can extend the life cycle of buildings and their components in a resource- – and cost- – efficient manner to house a growing population. In shifting perspectives to value and maintain precious resources, designers should not only apply existing local materials (in the form of raw, useless, by-products, hybrids, and/or offcuts) in innovative ways to support changing housing uses, needs, and environmental demands but also design and plan demolition differently and more efficiently.

The proposal for reducing demolition waste, raw resource demand, and emissions or, ultimately, eliminating construction is to re-use the materials already in circulation to avoid contributing to going beyond the planetary boundaries which "define the safe operating space for

DOI: 10.4324/9781032665559-18
This chapter has been made available under a CC-BY-NC-ND 4.0 license.

humanity with respect to the Earth system and are associated with the planet's biophysical subsystems" (Rockström et al., 2009). An overall shift in practice to utilising previously used, by-products, and "waste" materials requires a greater cultural shift in mindset. This is the exact subject of our two-week programme where we were joined by students from all over the world.

The results are place-based, responding to the local environmental conditions as well as the availability of materials and labour. This was evidenced by the variation in approaches and results produced by the diverse group of students who participated in the School of Re-Construction. As an online course, where students were able to join virtually from across six countries and two continents, the exercises of identifying the local housing issues and identifying the available material and skilled solutions produced a range of responses specific to the students' respective origins.

Methodology and learning outcomes

This summer school provided an opportunity for students to share contextually unique experiences and skills and expand their knowledge about re-use of materials, circularity, and development of positive and impactful projects. The selected students were mainly from an architecture background who were deeply aware of the current climate and ecological crisis, were mindful about the impact of the built environment in our planet, and equally felt the frustration of slow governmental change. Through active research to address problems and by looking into case study findings, the students were able to work collaboratively and engage with architects, tutors, and researchers.

The methods encouraged, from the collection of relevant precedents to the gathering of locally available references, allowed students to frame an approach through a system of diagrams where they could visualise methods of construction, materials, and refurbishment alternatives. Ultimately, the main goal would be to understand how to perform a pre-deconstruction audit of their own project whilst mapping materials of their chosen environment and building components available to them. The design proposal would then include a digital catalogue of components to provide a visual aid for discussion and a sketch catalogue. These individual contexts would respond to the appropriate themes (raw, useless, by-product, hybrid, offcut) and set a baseline for the final proposal.

Therefore, after learning basic definitions, attending online presentations, and putting into practice the tools and resources provided, the students were able to understand and articulate a holistic approach to architecture. The outcomes enabled them to recognise the benefits of circular economy and upcycling, material flows, and the incentives to increase material (and components) re-use with different stakeholders and at various scales. They were equally able to create long-term and short-term strategies to close the loop whilst applying systems thinking design in a comprehensive approach to housing design. In the end, they took home a replicable working method for future reference and application.

Programme – Week 1

Prior to the start of the course, students were encouraged to do some preparation for the seminars, discussions, and activities through reading and watching as a way of reflecting on the basis of circular economy. Students received a reference list with videos, documentaries, literature, and podcasts relevant to the topic of housing as a way of finding inspiration to the possibilities of applying circular economy principles to existing housing stock. Resources included the Architects Climate Action Network (ACAN) circular economy series, documentaries such as *Waste Land* by Vik Muniz and *San Wars* by Denis Delestrac, and literature from Lendager, Metabolic, and AECOM.

The first week kicked off with an insightful presentation about the basic terms and definitions of circularity and how to apply them to different stages of the process, using RIBA stages as an example outline but allowing it to be applied in different contexts. If you are familiar with the RIBA Plan of Works, you will recognise that Stages 0–7 do not have in consideration the work of evaluating existing structures and their components for deconstruction and potential re-use nor the demolition process. In the spring of 2021, ACAN proposed a new Stage 8 to close the loop on the RIBA working stages not only to address a flaw in the design process and construction stage but also to push architects to think beyond handover procedures. By providing examples of a potential Stage 8, we are raising awareness for the urgent need to change the current linear practices into a circular system that will allow the re-use of materials and reduce waste in construction and manufacturing.

In this first session, students were able to discover different types of housing and types of reconstruction, precedents, and case studies to be used as a starting point for their research. The students collected various examples from low-cost homeless shelters to zero-waste buildings, from Irish eco-villages to retrofits and were able to identify problems associated with housing and construction methods (see Figure 13.1).

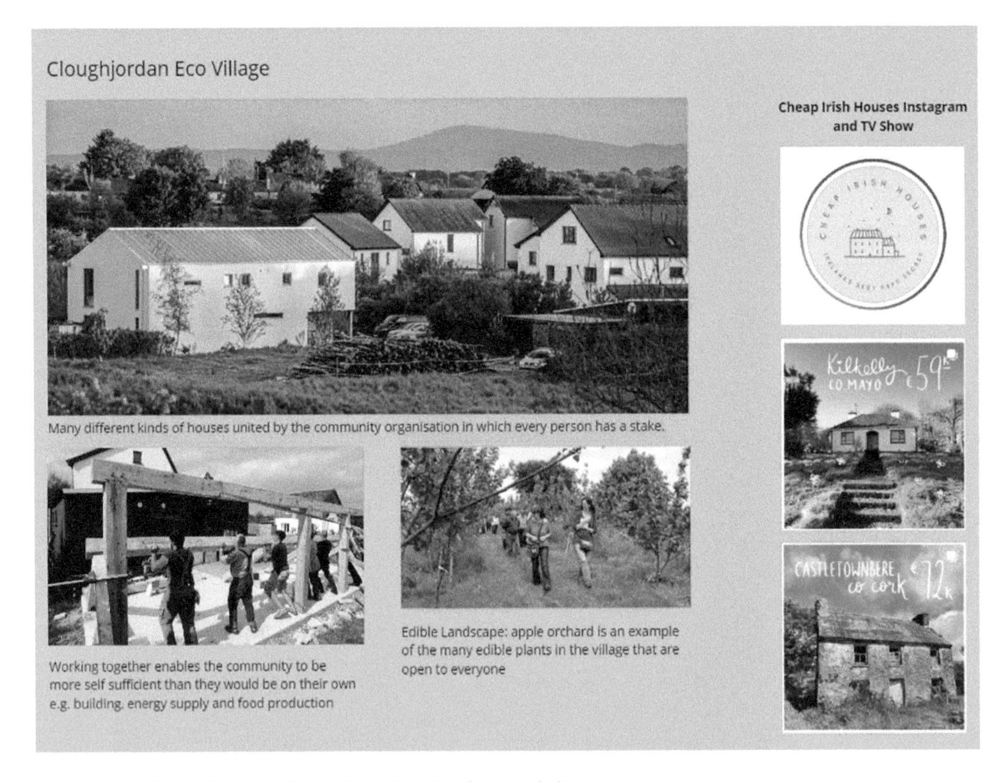

FIGURE 13.1 Precedents collected during the first activity

Source: Simon Schaubroeck, Luke Hardman, and Callum Purdue

Systems thinking

As a way of understanding and analysing complex systems, students were introduced to the concept of systems thinking to support a holistic approach to the work at hand (Schlüter et al., 2023). This allowed them to focus on the relationships and interactions between the various components of a system rather than just looking at individual parts in isolation. Systems thinking is essentially a system of thinking about systems (Arnold & Wade, 2015) where it recognizes that everything is connected and acknowledges that changes to a part of a system can have ripple effects throughout the entire system. It is important to consider the broader context and external influences which can impact the behaviour of the system as a whole (see Figure 13.2).

To understand the dynamic in a system, it is essential to highlight the importance of interconnectedness and interdependencies of the various elements within the system boundary (Flood & Jackson, 1991). As opposed to linear thinking, a system is a tool that can amplify, reinforce, or restrain relationships between elements and external factors or 'the environment'. The inputs, outputs, and feedback loops are the processes that occur within the system to achieve its purpose, and by visualising the elements and relationships between them, it helps identify the leverage points that can influence a system's outcome and promote a comprehensive approach to decision-making (Meadows, 1999).

The exercise consisted of a discussion through role-play, where students paired with each other to represent a stakeholder of the built environment: local authority, architect, consultant, contractor, client/occupant, etc. This allowed the students to challenge different points of view by improving communication and develop a strategic and systemic approach for collective problem-solving. Effectively, to facilitate the complexity of a system, students were encouraged to map loop diagrams (see Figure 13.1), identify root causes and leverage points (see Figure 13.2) against design criteria to study possible solutions.

These tools as well as this exercise proved to be an excellent resource to find the root cause of the problem discussed from the students' initial findings and define a case study. Once a clear and defined methodology was in place, the students were introduced to pre-demolition audits and the process of understanding the types and quantities of materials that are available and recommendations on how these materials can be managed.

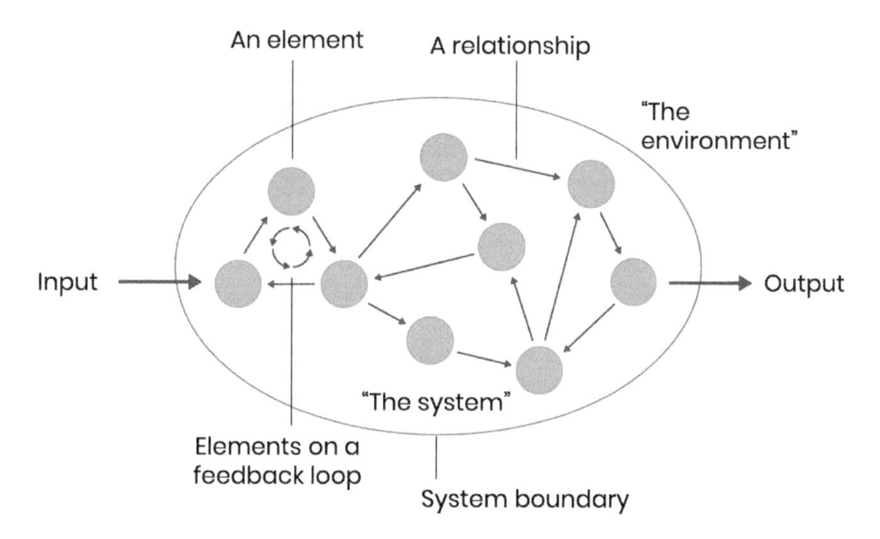

FIGURE 13.2 Diagram based on Flood and Jackson (1991)

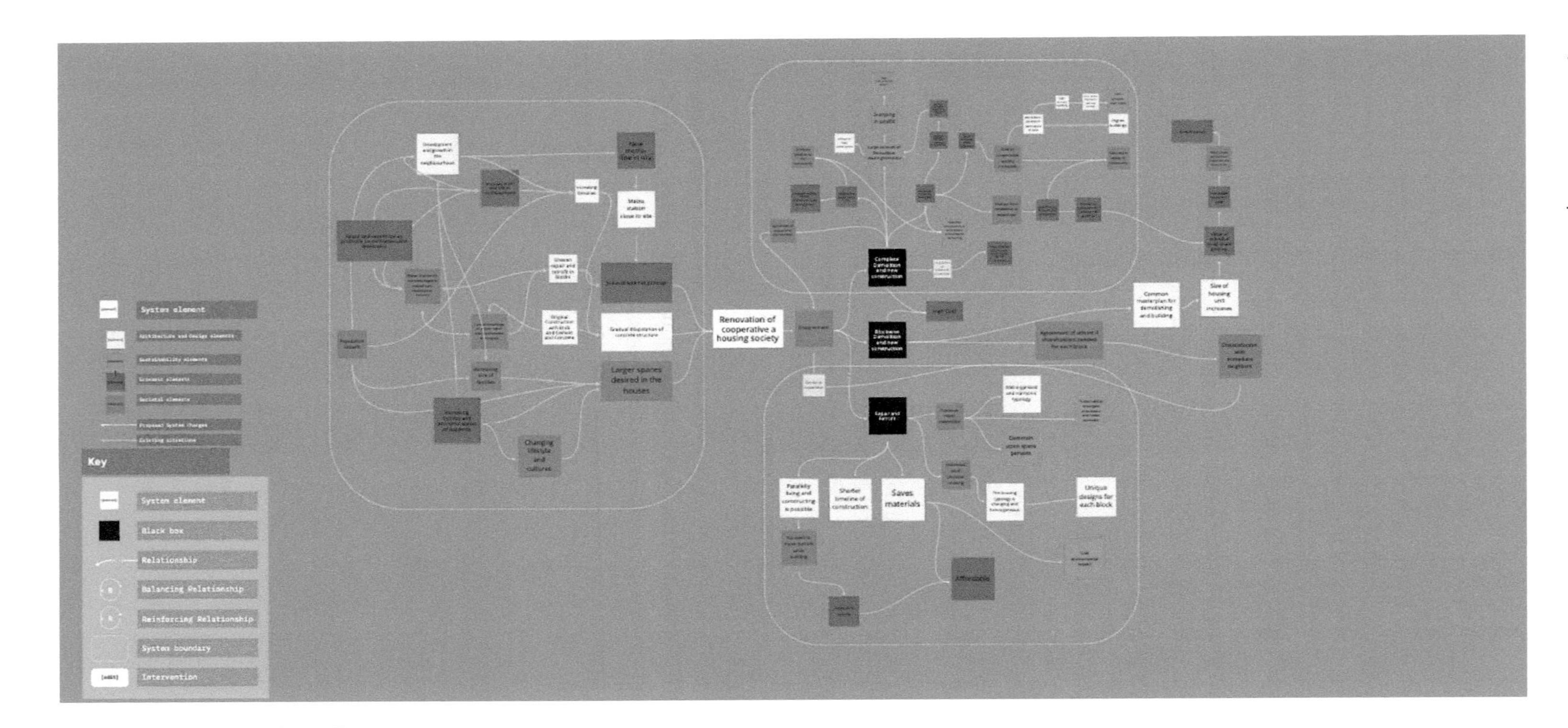

FIGURE 13.3A Causal loop diagram

Source: Divya Chand and Razan Atwi

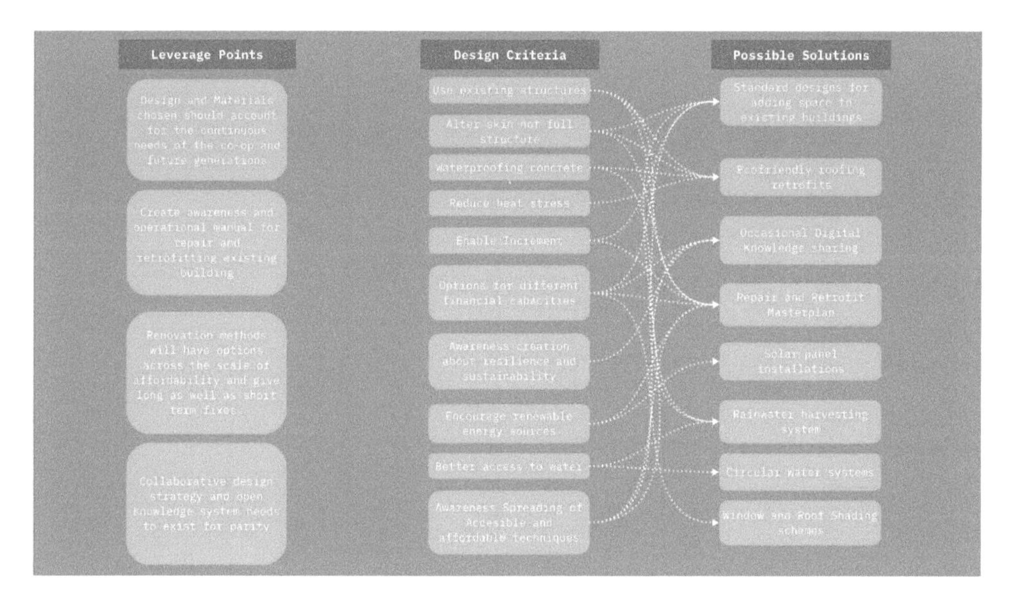

FIGURE 13.3B Leverage points, design criteria and possible solutions

Source: Divya Chand and Razan Atwi

Pre-demolition audits

> Pre-demolition and pre-refurbishment audits provide clients with independent advice about the products and materials that can be reused or recycled prior to demolition or major refurbishment. The client can then use this report to set targets and objectives within the demolition or refurbishment tender documents to ensure best practice in resource efficiency is adopted by the appointed contractor.
>
> *BRE Group definition of pre-demolition audit stated on the BRE Group website*

When considering an existing building and its demolition, the best practice is to assess the building elements and materials and their potential value in order to fully understand which approach to take (refurbish, retrofit, demolition). By identifying the building materials that can be recovered and products presenting a high re-use potential, this results in a 'reclamation inventory', where building materials can be identified and ranked. The resulting inventories showed the materials' and products' characteristics, such as dimensions, quantities, conditions, environmental impact, technical characteristics, and disassembly recommendations (see Figure 13.4).

Through analysing the material stock and quantifying it, it was possible to evaluate the challenges and the potential of the materials surveyed. The students were able to develop their proposals based on the resources and tools provided in the first week.

For a better understanding of how it works, we have attached a pre-deconstruction audit template and materials catalogue that can be used for the following exercise. The resulting materials catalogue will inform the materials you have available to you for the re-design/re-construction process.

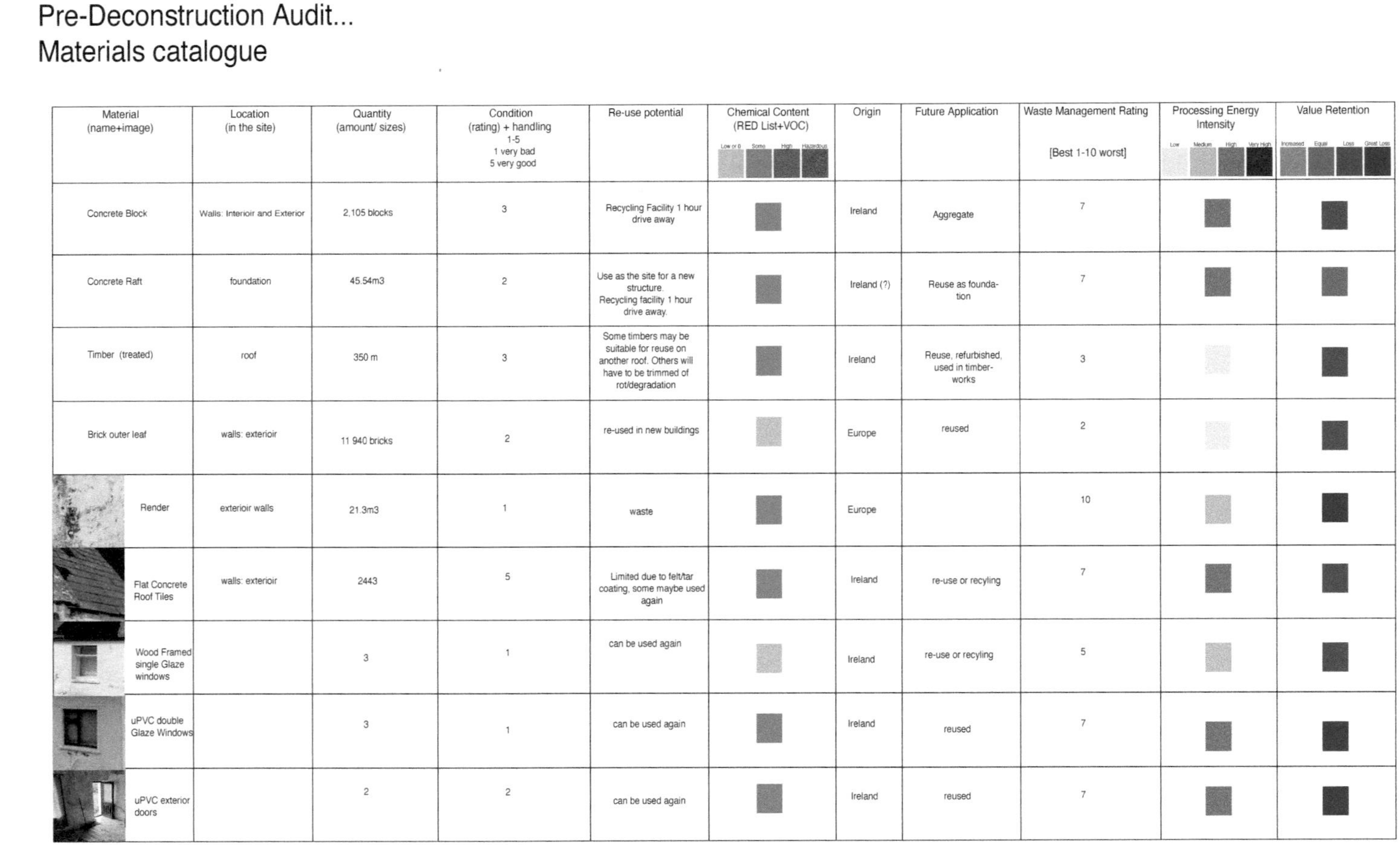

Pre-Deconstruction Audit...
Materials catalogue

Material (name+image)	Location (in the site)	Quantity (amount/ sizes)	Condition (rating) + handling 1-5 1 very bad 5 very good	Re-use potential	Chemical Content (RED List+VOC) Low or 0 / Some / High / Hazardous	Origin	Future Application	Waste Management Rating [Best 1-10 worst]	Processing Energy Intensity Low / Medium / High / Very High	Value Retention Increased / Equal / Loss / Great Loss
Concrete Block	Walls: Interioir and Exterior	2,105 blocks	3	Recycling Facility 1 hour drive away		Ireland	Aggregate	7		
Concrete Raft	foundation	45.54m3	2	Use as the site for a new structure. Recycling facility 1 hour drive away.		Ireland (?)	Reuse as founda-tion	7		
Timber (treated)	roof	350 m	3	Some timbers may be suitable for reuse on another roof. Others will have to be trimmed of rot/degradation		Ireland	Reuse, refurbished, used in timber-works	3		
Brick outer leaf	walls: exteriour	11 940 bricks	2	re-used in new buildings		Europe	reused	2		
Render	exterioir walls	21.3m3	1	waste		Europe		10		
Flat Concrete Roof Tiles	walls: exterioir	2443	5	Limited due to felt/tar coating, some maybe used again		Ireland	re-use or recyling	7		
Wood Framed single Glaze windows		3	1	can be used again		Ireland	re-use or recyling	5		
uPVC double Glaze Windows		3	1	can be used again		Ireland	reused	7		
uPVC exterior doors		2	2	can be used again		Ireland	reused	7		

FIGURE 13.4 Reclamation inventory/materials catalogue

Source: Callum Purdue and Jekaterina Ancane

Exercise

Using the leverage points and potential design solutions identified in the systems thinking exercise as a starting point, create a Local Resource Map & Material Inventory (with analysis) using the pre-deconstruction audit tools.

Consider 'useless', 'raw', 'off-cut' materials in addition to the obvious building materials available.

Location	*What is the most common construction type in your city?* *What buildings are being frequently demolished/replaced and where is it occurring?*
Material Type	*Which material is reusable? In the same place?* *Which material is recyclable? Impact of recycling on environment and cost?* *Which material would you enjoy building with?* *Which material has scalability potential?*
Extraction Process	*How do you maintain material quality?* *How do you store and recirculate the building material?* *How do you compete with virgin construction material?* *How do you ensure supply meets demand?*

Programme – Week 2

In the second half of the programme, limited tutoring was provided to assist with the students' proposals and a presentation prepared ahead of the group critique session. The final assessment included action plans, solutions to repair and retrofit, and water and waste management from all student groups (see Figure 13.5a and 13.5b).

FIGURE 13.5A Extract from the final presentation

Source: Divya Chand and Razan Atwi

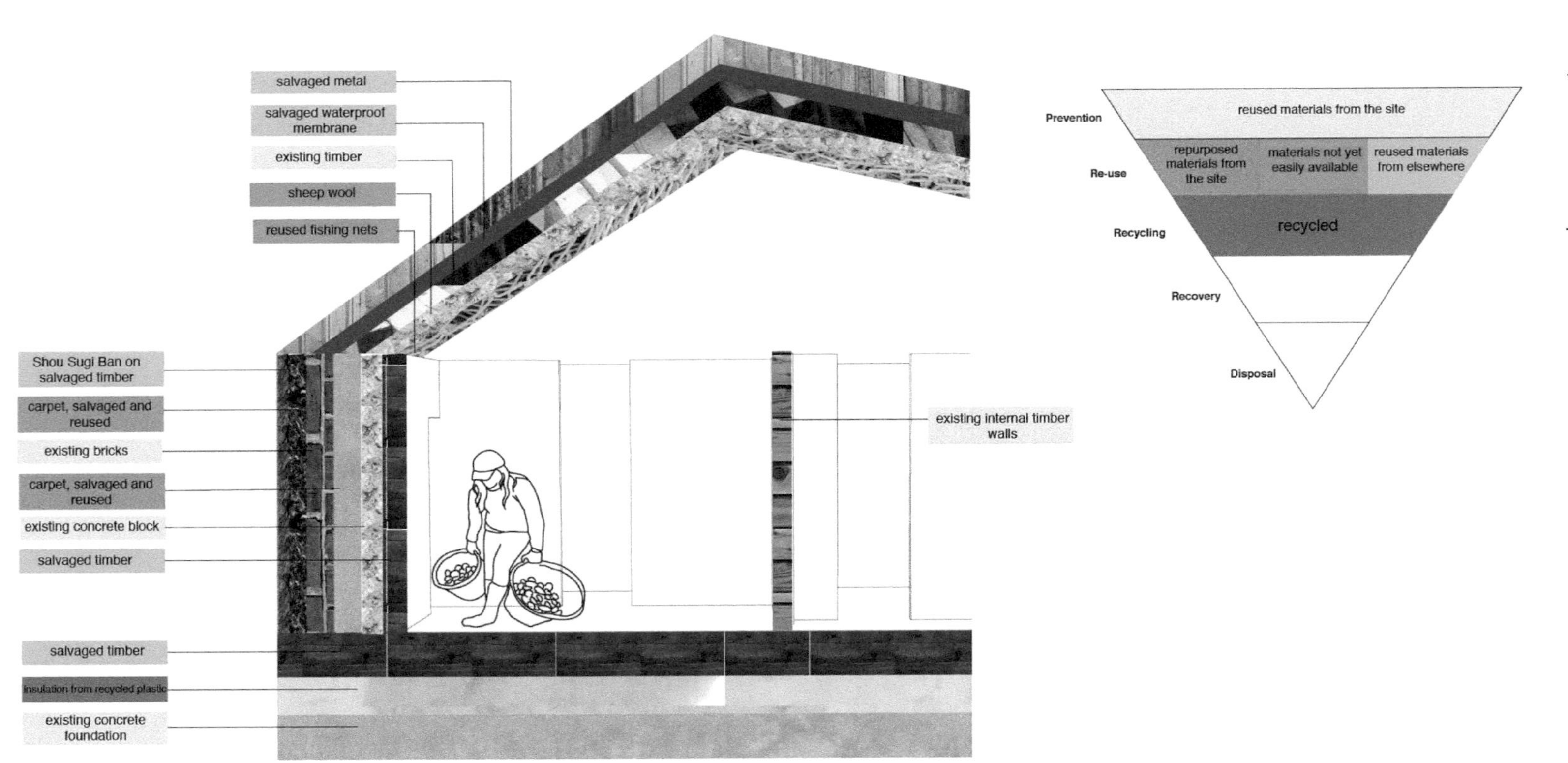

FIGURE 13.5B EXTRACT FROM THE FINAL PRESENTATION

Source: Callum Purdue and Jekaterina Ancane

The final reflections were extremely rich with ideas and content, and there was a genuine ambition to make work with the brief given and create small changes with a huge impact. The feedback from the students was rewarding (as stated at the end of this chapter), and they have applied the learnings to their academic paths and future careers.

Feedback from students

While I was already familiar with the basic principles of circular economy and overall benefits of upcycling, the two weeks of this summer-school were an extremely valuable learning experience for me. I enjoyed the opportunity to apply these, using the important tools of systems thinking and design, in a more practical sense to my case. As I come from an Urban Studies background, I especially appreciated the fact that I was allowed to explore the more socio-economic aspects of repair, reuse and reconstruction in the exercises, something I've rarely gotten to do within faculties of architecture.

The approach and tools of systems thinking I picked up during our exercises have come handy repeatedly in the last couple of years. On familiarising myself further with the literature shared, I've conducted workshops on the topic for colleagues and my students. The methods of root-cause maps and defining design criteria also came in quite handy for synthesising research findings, presenting leverage points systematically and developing project strategies with clients in the rural housing sector.

The deconstruction audit was also a very useful exercise for me. In a team with Ar. Razan Atwi, I had the opportunity to apply these learnings in a practical way to a case study that I have a personal connection with, the housing co-operative society I live in and grew up in, where discussions of reconstruction and redevelopment were brewing. The systems-thinking exercises in tandem with the pre-deconstruction audit we did reinforced the need for repair & retrofit solutions instead of demolition for our cooperative society, showing how much use-value can be retained in the existing materiality of the housing. I also delved into exploring the scrap-material flows and local resources in and around Nagpur through these exercises which led to some fascinating findings about the highly dense, informal networks and various scales of stakeholders involved. Since the Summer school, no solution for the demolition or repair of this neighbourhood has yet been arrived at and conflict continues as plans for redevelopment are brought up, pushed for and opposed in varying measures. For our particular housing unit, we conducted repair work on the RCC roof, ensuring heat proofing with clay-lamps sourced from a local market. These are hand thrown and usually used for religious purposes. My push to incorporate these into the roof was met with trepidation by the masons and contractors but SoR-C had me convinced and determined that it is an idea worth fighting for. For the flooring on the top-terrace, we sourced boxes of broken vitrified tiles from suppliers at 1/5th the cost, and I trained the local workers into doing a china-mosaic flooring. These are great at reflecting back the harsh sun, and give an extra layer of water-proofing to the house. Since implementation, the roof has withstood 2 harsh Indian summers and rainy monsoons and ensured the house stays cool and dry! It's a success, and the repaired house is testament to the quality and value of the existing project. While the other units, poorly maintained and un-repaired, seem fit for demolition, this equally old unit stands strong and keeps us comfortable. I continue to advocate these repair & retrofit solutions instead of demolition for our cooperative society, showing the residents how much use-value can be retained in the existing housing (and to make the case for deconstructing the skin and not the whole blocks). I have shared some photos of this project with this email.

In SoR-C, it was inspiring to see the diversity of speakers in the keynotes and panels, and the variety of ways of practising the same principles of circularity. I have gone back to the recording's multiple times for inspiration for graphics, illustration, case studies of not just projects but practices as well. Thank you Filipa and Taleen for those two days. Having since taken up teaching online, I have also gone back to how well structured our time and interactions during those 2 weeks were as reference to design my courses.

Divya Chand Partner, Lokal Habitat Labs LLP
India Smart Cities Fellow, MoHUA

I loved it. The only shame was it was online because of COVID. It really helped clarify issues I was worrying about in terms of global concerns and how I could use my career to tackle them in a meaningful way. And the experience of taking a systems approach to architecture was really interesting. Helped to bring in some of my previous studies in philosophy and sociology in a relevant way to architecture. I actually used a lot of the thinking in my final year studio project that I have just completed this year, which proposed a new grade 0 listing system for buildings and components that means they can't be thrown away, and have to be reused in some way instead. The idea was to convert an existing modernist car park and hotel into a working museum with heritage England and designers under one roof working on reuse projects and displaying them to the public. And a lot of that was feeding directly from the summer school on reconstruction! And it also just introduced me to such interesting source material and architect practices in general, which I'm really grateful for.

Callum Purdue Part 1 Student

References

Arnold, Ross & Wade, Jon. (2015). A Definition of Systems Thinking: A Systems Approach. *Procedia Computer Science*, 44, 669–678.

Beadle, K., Gibb, A., Austin, S., Fuster, A., & Madden, P. (2008). Adaptable futures: Sustainable aspects of adaptable buildings. In: Dainty, A (Ed.) *Procs 24th Annual ARCOM Conference*, 1–3 September 2008. Cardiff: Association of Researchers in Construction Management, 1125–1134.

Chuai, Xiaowei, Lu, Qinli, Huang, Xianjin, Gao, Runyi & Zhao, Rongqin. (2021). China's Construction Industry-Linked Economy-Resources-Environment Flow in International Trade. *Journal of Cleaner Production*, 278.

Flood, Robert & Jackson, Michael. (1991). *Creative Problem Solving: Total Systems Intervention*. Chichester: Wiley.

Meadows, D. (1999). *Leverage Points: Places to Intervene in a System*. Hartland, VT: The Sustainability Institute.

Rockström, J., Steffen, W., Noone, K., et al. (2009). A Safe Operating Space for Humanity. *Nature*, 461, 472–475. https://doi.org/10.1038/461472a

Schlüter, Leonie, Kørnøv, Lone, Mortensen, Lucia, Løkke, Søren, Storrs, Kasper, Lyhne, Ivar & Nors, Belinda. (2023). Sustainable Business Model Innovation: Design Guidelines for Integrating Systems Thinking Principles in Tools for Early-Stage Sustainability Assessment. *Journal of Cleaner Production*, 387, 135776.

PART IV

Reflections on re-use teaching and practice

14

HOW TO TEACH ARCHITECTURAL DESIGN IN THE (NEW) AGE OF CONTINGENCY?

Lionel Devlieger and Maarten Gielen

It is highly unlikely that today's students will still be able to see virgin steel and concrete as the go-to construction materials by the time they establish their own practices.

Societal concerns about environmental impact will necessarily imply using high-impact, energy-heavy resources more sparsely or for very long useful lives. The relationship between architects and materials will become more complex as the profession is increasingly asked for accountability on its tremendous environmental footprint. Materials that are ubiquitous today will need to be substituted by other types of materials: think of biobased materials such as lumber or local mineral materials such as clay. And let us not forget the heterogenous category that is of particular interest to Rotor (DC): materials and components from earlier constructions, harvested and prepared for re-use in new projects.

The re-use of building materials leads to immediate and sizeable savings in environmental impact. Sooner rather than later either legislators and/or rising costs will force the building industry to rethink its material sourcing.

Such a radically different use of materials will profoundly change the architectural profession in the coming decades, just like the emergence of reinforced concrete reshaped the building sector and its protagonists in the 20th century. This constitutes an important challenge for those in charge of educating future professionals. As the old world of cheap oil and coal is dying, the new circular world struggles to be born. What professional perspectives can we give to students today? Will our societies tear down and build as much as we do today? Will the design tools we use today still be relevant? What impact will come from the skyrocketing prices of certain materials? Being honest about those uncertainties is a prerequisite for becoming a trustworthy teacher. But, what then is there to teach?

First we need to do away with the dichotomy between two stereotypical teaching methods. A first approach sees architectural education strictly as professional training. In guiding the student's work, the instructor simulates the contingencies of architectural design, that is, the constraints imposed by a wide range of parties the student is likely to negotiate with in his/her later professional life in order to provide a 'realistic' experience to the student. An obvious critique on such an approach is that this imprisons the profession. Architects are at the mercy of

DOI: 10.4324/9781032665559-20
This chapter has been made available under a CC-BY-NC-ND 4.0 license.

FIGURE 14.1 Student Samuel Little (Dip.18, AA School, London, 2018/19) has created an imaginable scenario: a trader of surplus steel sets up another company that converts salvaged steel parts into prefabricated portal frames. A partner company followed his suggestion and invested in the idea.

their commissioners, legislation and the market economy during their entire career. If not during their studies, when will they have the breathing space to come up with alternative ideas? And, moreover, it is to be questioned whether the teacher can successfully simulate lifelike, 'realistic' contingencies.

On the other side, there is the idea that architectural training should offer unlimited freedom and push the student to the unexplored limits of his/her imagination. 'Design a house on Mars', 'suspend a neighbourhood on helium balloons', 'design a building inspired by broccoli' etc. While such absolute – artistic – licence would certainly be therapeutic for architects with a few years of professional experience, it is often not beneficial for students. Uncertain about exactly what they have been given freedom from, many students turn to the 'oeuvre' of the design studio instructor in a desperate attempt to find some form of constraint that can inspire their work.

How do you overcome the impossible choice between those two extremes? An extra ingredient is needed in the discussion: the crucial idea that, not only physically, but also mentally we live in constructs of our own making: institutions, legislation, disciplinary frameworks and even our economy and political organigrams are entirely artificial. Hence, these can be challenged and even changed. Our students' choice is then no longer limited to either accepting the constraints of the architectural profession or hiding from them in the safety of formal explorations. They are invited to invest a wider realm with their talent to redesign things and to come up with a plausible proposal on how existing institutions can be reimagined and what the role of the architectural profession should be in the new age of contingency.

MAP 14.1 Screenshot of Beyond Opalis website – World map of locations of Opalis secondhand materials' suppliers as listed in their directory

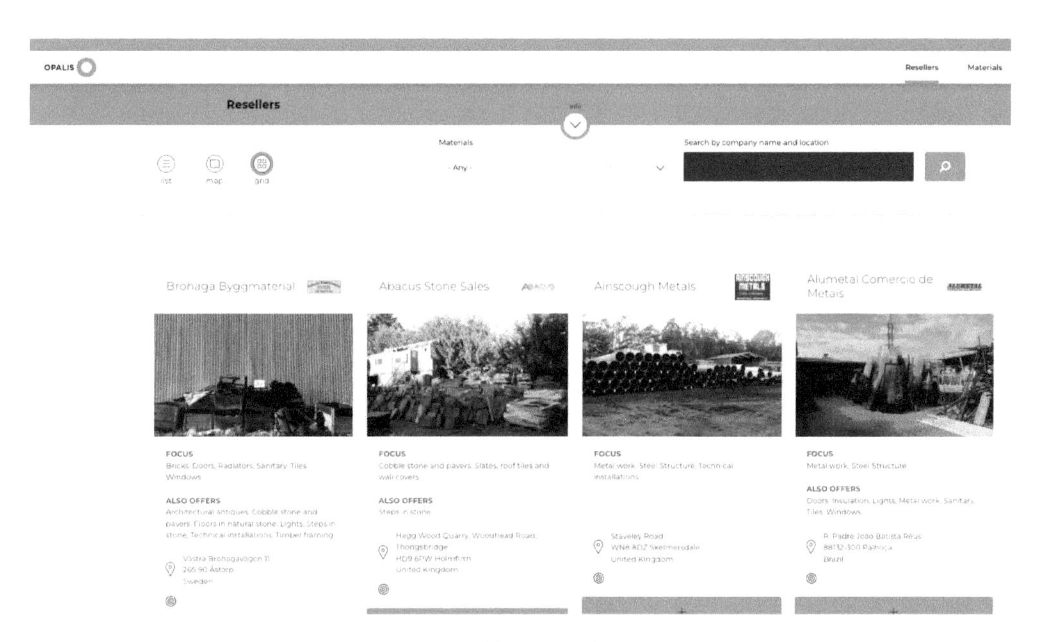

FIGURE 14.2 Beyond Opalis is a student-run directory of professional dealers in secondhand building materials. The project was initiated in 2018 when Rotor was teaching the Studio 18 diploma course at the Architectural Association (AA) School of Architecture in London. Since then, the directory has been maintained and expanded internationally. Soon, it will be integrated in the main website under the name 'Opalis Sandbox'.

FIGURE 14.3 Student Amaya Hernandez (Dip.18, AA School London, 2021) discovered large quantities of re-usable materials while investigating buildings for demolition in central London. She organised their re-use and put a demolition company and a trader in touch with each other.

Source: Salvage of York stone pavement

FIGURE 14.4 The Master's Design Studio 'Challenges to Metabolic Design' at the University of Ghent (2021) sought solutions to challenges of circular building, such as logistics issues. Gentiel Acar, Jesse Ghyssaert, Ferre Lust and Karen Steukers designed a multifunctional centre for the production, processing and sale of reclaimed materials. They designed the building from existing steel portal frames.

Addressing the re-use of building materials provides a good alibi for such systemic thinking. The absence of a market in perfectly predictable grades of materials forces the students to address the question of materials much earlier in the design process. The current economic conditions that favour cheap new materials over expensive labour need to be reimagined. Building prescriptions and esthetical preferences need to be weighted on their merits. And throughout these processes, the relationship between design and execution takes on new forms, while new and surprising alliances between architects and other practitioners see the light.

That some of the students' proposals reach a receptive audience beyond the school shows that there is a willingness in the building industry to embrace new ideas – and also a dire need for spaces where they can come up.

Originally published in

Bauwelt n°233 (2022) p. 64–65, as "Wie kann man Architektur im (neuen) Zeitalter der Unwägbarkeiten lehren? Wiederverwertungsstrategien im studentischen Entwurf"

15

REVERSIBLE BUILDING DESIGN STUDIOS DURING GREEN DESIGN BIENNALE

Elma Durmišević

Introduction

At the core of all design concepts and interventions in the built environment lays the question: How can interventions in the built environment eliminate negative impacts on ecological system and biocapacity of the planet and transform them into a positive once? *Circularity Gap Report* argues that to bring human activities back within the safe limits of the planet, global material extraction and consumption would have to be reduced by one-third (*Circularity Gap Report*, 2022). At the same time, the report from 2021 indicated that almost 60% of the built environment required to accommodate urban population by 2050 remains to be built (*Circularity Gap Report*, 2021). This gap can be bridged by multiple and effective re-use of resources and industrial concepts where waste does not exist and where materials from one process are resources for another. This chapter presents a new design concept (applied during students workshops in the past six years) which can unlock multilayered re-use capacity of buildings and their components/materials and enable their multiple re-use options. Such approach envisions the built environment without demolition and value degradation and labels construction and demolition waste as design error. In order to eliminate this design mistake from the design process International Green Design Biennale organizes regular multidisciplinary design studios involving students from all over the world in exploring new dimensions of the future generation of buildings – reversible buildings.

International Design Studio results presented in this chapter have addressed two key elements of reversible buildings: (i) spatial reversibility (accommodate multiple functions and spatial configurations without demolition) and (ii) technical reversibility (consider individual recovery and exchangeability of building elements for re-use). Spatial reversibility defines the ratio between fixed and variable space and the capacity of variable space to accommodate different functions while technical reversibility focusses on key indicators of technical reversibly such as independency and exchangeability.

Together, spatial and technical reversibility define the circularity profile of a building. Buildings with high spatial end technical reversibility are true circular buildings, enabling long use life of the building by smart design of fixed parts of the building while providing capacity to

DOI: 10.4324/9781032665559-21
This chapter has been made available under a CC-BY-NC-ND 4.0 license.

adopt built structure to individual needs leaning on technical reversibility of variable parts. Having this in mind, a correct definition of a circular building would be "circular building equals flexible monument", which has been partly demonstrated through design studios during Green Design Biennales.

Green Design Biennale

For the last 10 years, Green Design Biennale marks October days in Mostar, Bosnia and Herzegovina. Green Design Biennale is a dynamic multidisciplinary platform that every two years brings together designers, architects, urban planners, researchers, PhD students, Information and Communication Technology (ICT) experts and engineers from all over the world, with the aim of promoting new design principles and solutions that could help to eliminate adverse impacts of modern civilization on the ecological systems of the planet. Special focus of the Biennale is on the digitalization and implementation of circular economy in architecture and the design of healthy and inclusive green buildings and cities. A number of exhibitions, lectures, town hall meetings, green fashion show and student workshops involving on average of five universities per biennale year are organised to highlight these themes.

As initiator and curator of Green Design Biennale, my aim was and still is to work towards strengthening awareness and scientific capacity in the transition towards circular economy and focussing on presenting green and circular city/building solutions and educational advancement in the field of reversible building design, which has been a center point of international design studios during each Green Design Biennale. In the course of the last 10 years, Biennale appealed to many supporters starting from United Nations Development Programme, United States Agency for International Development and Swedish International Development Cooperation Agency, European Union, Architects Council of Europe, Dutch embassy, Cradle to Cradle, Delft Technical University, University of Munich, Istanbul

FIGURE 15.1 Overview of activities and students work during Mostar Green Design Biennale 2022

Source: https://sarajevogreendesign.com/sgd2022/

Technical University, University of Sarajevo, University of Dzemal Bjedic, University of Mostar and many others.

Reversible building design

Reversible building is a backbone of circular building and a key enabler of the circular economy in construction. It is a driving force behind circularity of building and its materials and their potential multiple applications in the future. *Reversibility* is defined as a process of transforming buildings and dismantling, recovering and re-using the systems, components and elements without causing damage to the building components and elements. Building design that can support such processes is reversible (circular) building design (E. Durmisevic 2018).[1]

In other words, reversible building design is design which enables buildings to be i) modified to meet different use or technical requirements without demolition and ii) enables building components/elements to be recovered without damaging surrounding products/elements and be reconfigured/upgraded/repaired and re-used in a new building or system (E. Durmisevic 2020).[2]

A reversible building design framework including design guidelines, protocol and tools to assess reversibility has been tested and validated during the H2020 Buildings as Material Banks project. This chapter will illustrate how this concept has been explored through multidisciplinary international design studios with students from seven EU universities including students of architecture, industrial design and civil engineering. International design studios have explored building design parameters that address spatial capacity and technical reversibility of building

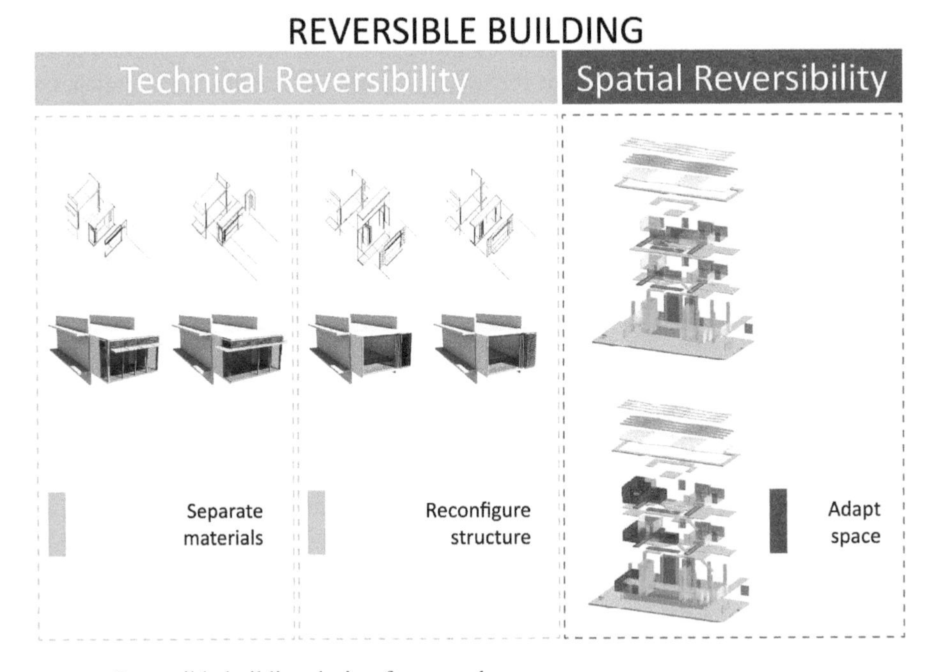

FIGURE 15.2 Reversible building design framework

Source: Elma Durmišević 2018 (www.bamb2020.eu/wp-content/uploads/2018/12/Reversible-Building-Design-guidelines-and-protocol.pdf)

as well as how to design new buildings while reusing more than 70% of existing building product and element.

Reversible building design studio brief 1: reversible urban pixel

This design studio focussed on the design and prototype of a *reversible urban pixel (Pixel)*, which can be transformed to meet three use scenarios. The reversible urban pixel is envisioned as a space that can be placed in different urban areas while adjusting its form and function to the needs of the particular spot. Students had to use existing elements from two existing building systems in designing a Pixel structure for reconfiguration and reassembly so that elements can circulate from one use scenario into another. Fifty students from different disciplines investigated complexity behind the concept of re-using existing building elements.

De Groot Vroomshoop (a construction company from the Netherlands) provided the wood façade subsystem and JANSEN (a steel supplier) provided two steel façade systems.

Design brief

The design task was to design a reversible system (transformable structure with high re-use potential of its elements) accommodating three functions and re-using at least 70% of the existing materials. The existing material bank consists of two types of wooden façade elements and steel profiles and components. Design needed to be reversible, meaning that the Pixel's configuration and its set of elements should be applicable in three use scenarios 'Working Pixel', 'Relaxation Pixel' and 'Commercial Pixel'. Each use scenario has to be (easy) reversible to the second and the third scenario while re-using materials from the previous one. Reversibility is not only introduced for these three use options but also for the future life beyond the three options. Besides an Urban Pixel needs to be self-sustained in terms of energy production for workstations and light.

International design studio methodology

At first, students worked on design of spatial concepts that can accommodate 'Working Pixel', 'Relaxation Pixel' and 'Commercial Pixel' and integrate multiple criteria regarding special and structural capacity, energy efficiency, comfort and design quality.

During this stage, students used the Multi Criteria Design Matrix (source E. Durmisevic 2016[3]) to identify design aspects, set up design priorities and, ultimately, evaluate their individual design solutions while searching for the most optimal integration and criteria trade-offs for the design brief.

Figure 15.4 illustrates the evaluation of three design concepts using the Multi Criteria Design Matrix as well as a presentation of the concept with the best evaluation score.

Once the spatial reversibility dimension was sorted out, the second work phase focussed on the design of technical reversibility. The objective of the assignment was to prototype the connections that make the 'reversible urban pixel' possible by bringing different elements of existing systems together. Extruded profile provided by JANSEN GROUP has been used as a base of the Pixels. During this design phase, students elaborated reversibility of different connection typologies that can meet reversibility requirements.

In order to demonstrate technical reversibility, the main focus was on investigation of constructive stability, optimalisation of the simplicity of the assembly and disassembly, damage

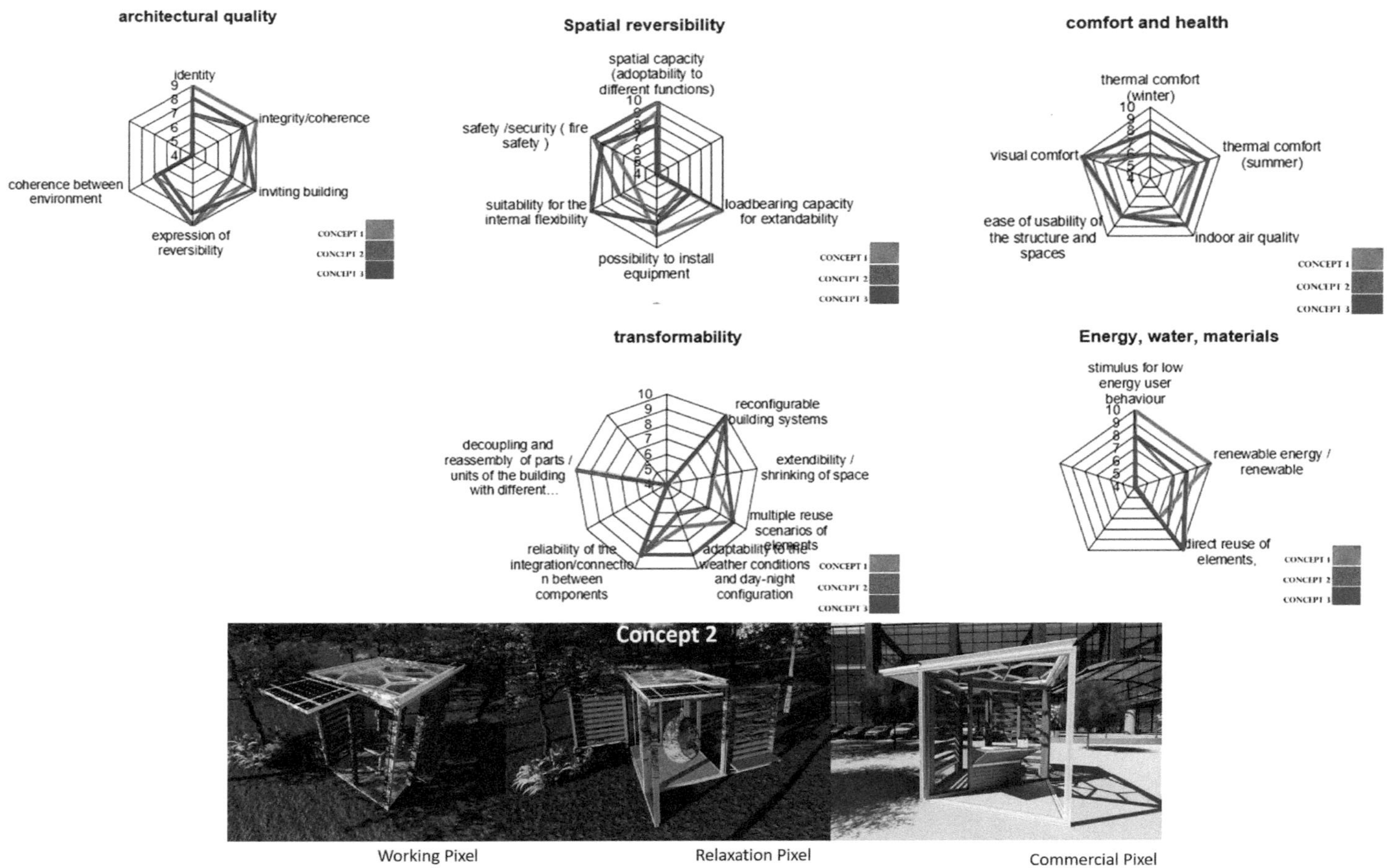

FIGURE 15.3 Multi Criteria Design Matrix and concept with the best score

sensitivity (how often can the connection be assembled and disassembled without losing its quality?), aesthetics of the connection and dimensional tolerances (what dimensional tolerances are necessary to increase the re-usability in new/existing buildings?). In total, eight aspects as defined next have been analysed for each proposed connection typology and solution.

Constructive strength

How much force structure and connection can take and when will it collapse. The final calculations haven't been made for different structural and connection types.

- **Dimension adjustability**

This refers to how flexible a connection is in terms of its dimensions. Is it possible to add another beam to the connection and can the structure expand or decrease in dimension simply by making a slight adjustment on the connection?

- **Re-use of materials**

This describes the re-use of existing materials to make the connection. A material that is added that can only be used for one particular connection scores lower than a material that can be re-used in other situations.

- **Damaging the structural beam**

This describes the amount of changes (damage) that come with a particular kind of connection. If one connection asks for lots of changes in the structural beam, it scores lower. For example, is it needed to drill or place a screw into the beams.

- **Capacity of the structural beam**

This subject describes the amount of room and possibilities there are to reinforce a beam while a particular kind of connection is being used.

- **Slot availability for façade elements**

This describes the amount of room and possibilities there are to connect any façade system to the exterior slots of the steel profile. If the main connection of the beam is using this space to connect the beams and columns, there would be no space left to make the connection between the structure and the façade system or panels.

- **Aesthetics**

This describes the architectural quality that the connections have. Internal connections usually score higher in this subject because they're simply not visible from the outside of the Pixel. Exterior connections, on the other hand, have a lower score because the connection will be visible from the outside.

- **Assembling speed**

This describes the difficulty level and time needed to assemble a particular connection. This subject is focussed on the assembling speed during construction and not during production. How many different steps and time are needed to make a particular connection while under construction?

In addition to the elaboration of suitable connection typologies, students have created a Revit model of the reversible urban pixel as a digital twin. This creates the opportunity to pass on the information from one building and re-use elements in a new building by their timely integration in design. Thanks to the Revit-based twin, the geometry and dimensions of materials could be directly used to model a new structure.

Evaluation wokshop took place in collaboration with the manufacturing company JANSEN GROUP. Scoring of individual connection types has been elaborated (Figure 15.4) and the two connection typologies have been merged into a final solution (Figure 15.4). JANSEN GROUP was very positive about the future potential of solutions and helped produce a prototype of

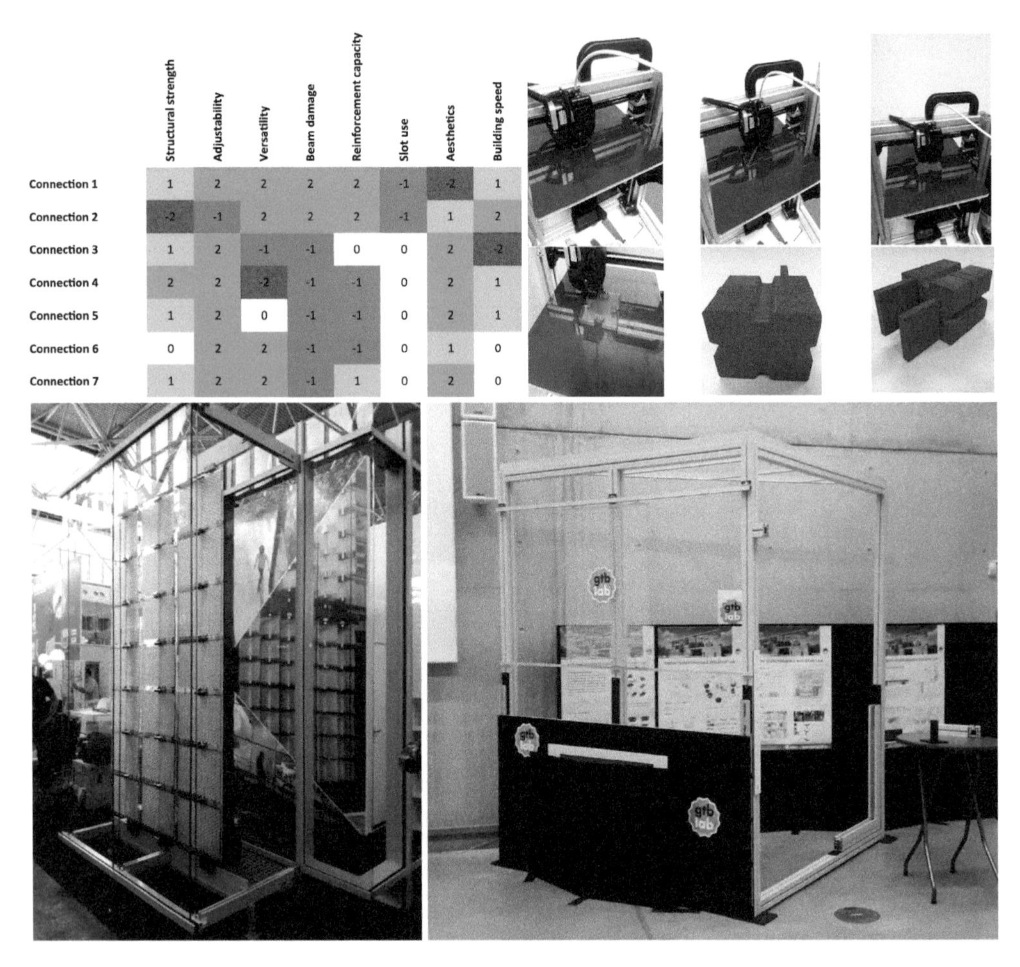

	Structural strength	Adjustability	Versatility	Beam damage	Reinforcement capacity	Slot use	Aesthetics	Building speed
Connection 1	1	2	2	2	2	-1	-2	1
Connection 2	-2	-1	2	2	2	-1	1	2
Connection 3	1	2	-1	-1	0	0	2	-2
Connection 4	2	2	-2	-1	-1	0	2	1
Connection 5	1	2	0	-1	-1	0	2	1
Connection 6	0	2	2	-1	-1	0	1	0
Connection 7	1	2	2	-1	1	0	2	0

FIGURE 15.4 Scoring of different connection types, prototype of reversible connection and simplifed Pixel demonstrating different connection possibilities and potential for reversibility

the final reversible connection type and demonstrated it in simplifed Pixel as illustrated in Figure 15.4c.

Reversible building design studio brief 2: reversible urban furniture

The main objective of the task given to the students of the international design studio during Green Design Biennale 2019 and 2022 was to design reversible furniture for the Green Design Centre (GDC) in Mostar using remaining material from its construction according to reversible building design principles.

Students had one day to visit to the construction site and hear more about the concept of re-using historic building as a base for new construction while unlocking cultural continuity and re-use of valuable parts from the past to build up a nucleus for innovation and creativity that will form a bridge to the future generation of buildings. The development of GDC presents a re-use of old ruin as a platform for construction of dynamic and exchangeable modern units demonstrating new approaches in design that enable disassembly, transformation and reuse of all its parts.

GDC was developed with local stakeholders bringing together steel manufacture, and wood cluster as well as local installation components around a reversible design and construction concept.

Students inspected remaining materials from the construction of the first phase of the Green Design Centre and had a week to create a design and produce a piece of furniture with the help of local manufacturers.

The first group of students was inspired by the form of the openings in the original stone fence and decided to use that shape as a connection point between the design of a reversible lamp and the existing building while re-using remaining translucent material (polycarbonate) and wooden panels and applying interlock connections without additional fixing devices.

A second group of students decided to use remaining wooden elements and panes in combination with translucent polycarbonate to create transformable furniture that is used as a bookshelf when positioned vertically with the lamp on top. When placed horizontally, the bookshelf becomes seating for three people with a lamp. They managed, in this case, to design and prototype transformable object while using gravity and interlocked connections without any additional connection devices.

FIGURE 15.5 Green Design Centre's materials being a material bank for design of reversible furniture

FIGURE 15.6 Design and process of making reversible lamp for Green Design Centre

FIGURE 15.7 Design and process of making reversible book shelf and seating for Green Design Centre

Summary

Students work presented in this chapter implemented dimensions of reversible building design while addressing i) multicriteria optimisation, ii) spatial and iii) technical aspects of reversible design, iv) the role of connectors and intermediary elements in unlocking reversibility and separation of martials without damages, v) how digitalisation can support circularity of re-usable elements, and vi) making and vii) transforming of reversible products.

Over 70 students with different disciplines (architecture, structural design, industrial design, civil engineering and interior design) from six universities Istanbul Technical University, Zuyd University of Applied Science, University of Mostar, University of Dzemal Bjedic, University of Twente and Sarajevo Green Design Foundation including support of Architects Council of Europe and Architectural Dialog Association came together to investigate and understand better design and construction challenges behind design for disassembly, design for reassembly, design for transformation and re-use of existing building materials in the development of new structures through design and prototyping of their design solutions and real life interactions with the manufacturing industry. This gave them a better understanding of challenges ahead that the industry and the whole ecosystem around resource circularity is facing, but they were equipped with tools and methods that can help them deal with future decision-making in a structured and systematic way.

Notes

1 Durmisevic E, *Reversible Building Design H2020 BAMB*, page 2, www.bamb2020.eu/wp-content/uploads/2018/12/Reversible-Building-Design-guidelines-and-protocol.pdf
2 Durmisevic E, *Design Strategies for Reversible Buildings H2020 BAMB*, page 40, Reversible-Building-Design-Strateges.pdf (bamb2020.eu)
3 Durmisevic E, International Design Studio 2010, Multicriteria Decision Making Matrix, Printed in the Netherlands ISBN: 978-90-365-3060-6

Reference List

Durmisevic, E., *Reversible Building Design Guidelines and Protocol*, page 2, EU BAMB 2020, published by University of Twente, 2018
Durmisevic, E., *Design Strategies for Reversible Buildings*, page 40, EU BAMB 2020, published by University of Twente, 2019
Durmisevic, E., *Green Design Centre Mostar*, published by NW Digital Deconstruction, 2021
Sarajevo Green Design Foundation, *International Design Studio, Green Design Centre*, published by NW Digital Deconstruction, 2021

16

RE-USE PEDAGOGIES

A reflection

Graeme Brooker and Duncan Baker-Brown

There has never been a more important time to teach how to re-use. That is, to re-use anything and *everything*. That said, where are we in the evolution and development of the pedagogies of re-using? This endnote is titled 'A reflection' because of just that, who is teaching re-use, how are they doing it and where can we see it?

Who

The majority of the current generation of prominent educators and designers of the built environment will have been taught in a manner that would not have dealt explicitly with the adaptation of existing buildings, cities and objects. Our (the authors of this endnote) generation had educational experiences that, on the whole, were blissfully free of the oncoming surge of anxieties surrounding our existence and the depleted resourcing of our beleaguered planet. I (Graeme) had an education at bachelor's and master's levels in interior design. Arguably all students ever did was work with existing buildings, but it was never explicitly defined or articulated as such until the master's experience, where words such as *context, continuity* and *site* became the new ways of explaining the existing. Material discussions and their extraction and subsequent distribution were predominantly about their aesthetic and occasionally cost manifestations. The idea of a material passport would have been met with blank looks. For me (Graeme), it wasn't until a secondhand copy of Victor Papanek's *Design for the Real World* and later during the master's thesis work that Philipe Robert's *Adaptations: New Uses for Old Buildings* and Stewart Brand's *How Buildings Learn* found their way into my possession that a future pathway was unlocked and the work of re-use came into sharp focus. Over 15 books and numerous articles and talks later, with thousands of bachelor's and master's students now educated in re-use, I can safely say the subject is still providing me with an undiminished intellectual stimulation and engagement aligned with an increasing anxiety of the foolishness of ignoring its basic tenets.

In stark contrast, today, any student on an architecture, interiors or any creative subject would have to work very hard to not be challenged and engaged with re-use, adaptation, issues around climate change, social justice, material depletion, resource extraction and so on. It would be unheard of to not be familiar with these terms. Yet, students will most likely be taught by the

DOI: 10.4324/9781032665559-22
This chapter has been made available under a CC-BY-NC-ND 4.0 license.

people of our generation as described before – people not schooled in these worlds but now tasked with inspiring others to be engaged.

So, how do we learn together? How can different generations of learners join together to understand and formulate ways of teaching future generations that design has fundamentally changed? It is clear that education has changed from outdated models of 20th century dogmas, filled with tabula-rasa planning, unfettered sites, purism, functionalism, rationalism (any 'isms' of universalising truths about design). Instead, re-use education will teach about the redistributions of material, wealth, justices, diversities, appropriations and authorships, copying and unending uneven resources availability, extraction, depletion. Pedagogies in design education are about scarcity, depletion, extraction, the fact that the designer is just one voice in a coalition of work, handed down, re-using. Now, how are these subjects taught, explored and developed, especially when people teaching and practising in them have not necessarily been schooled through them in their own formative educational experiences?

For us as authors, the School of Re-Construction workshop and the pedagogies of re-use are processes, methods and tools that are explored and will be explored, time and time again because they have yet to be fully outlined and articulated. When the School of Re-Construction was initiated, there were many discussions on who would take part, where were they, and what aspect of re-use, circularity and so on would they be assigned to. But, in this overview of participants, it became apparent that whilst numerous advocates could be located, there was very little written on re-use pedagogy. In the last two decades, much has been written about building re-use, circularity and the chronic and systematic challenges that we face as specialists of the built environment, but there is virtually nothing to be found on the education of designers in these processes and approaches. There are, undoubtedly, many hotspots of activity in numerous universities, practices and schools around the world, both in the Global North and South that are advocating re-use approaches. There are numerous agents tasking and testing their students with these processes. There are huge amounts of fabulous practice work, projects, unique initiatives in the reworking of our existing worlds, but teaching these processes and setting out the varieties of pedagogic tools with which to approach them are not yet coherently expressed. Why is this?

It is partly down to the reasons expressed before, that is, the shifting gears of generations and their responses to the world, but it is also a systemic failure of modern design education. Current pedagogies in design still prioritise the new over re-use. They still encourage the designer, the sole-designer at that, as the originator of the unfettered new invention/innovation. And this is a 'solution' that preferably has never been done before. Plagiarism still haunts the corridors of the institution, a situation which undoubtedly restricts pedagogic exploration. Yet, to copy out an idea, to roll it out repeatedly, and to utilise it again and again might just be the only way forward. I am reminded of an article written by Phineas Harper with regards to the Assembles Cineroleum project, the use of a deserted garage forecourt as a temporary cinema. The possibilities of rolling it out across the 4,000 or so deserted garage forecourts across the UK as a designed solution for all of the redundant spaces was poo-pooed as 'copycat', a sacrilege. Harper says,

> Were another practice to now create a version of the cinema in another town, they'd instantly be shot down as rip-offs or wannabes. Even Assemble couldn't create another Cineroleum without accusations of being lazy one-trick ponies. The exposure with which we reward good ideas in architecture is maddeningly also the death knell for those ideas catching on.[1]

In architecture, our comparable obsession with uniqueness drives the endless reinvention of form. Our lust for originality is wrecking the city, delivering a rash of formally new but ultimately anti-urban hideous skyline baubles reducing city-making to a spectacle of super-size billboard branding gestures while inhibiting the multiplication of good ideas. We must ask ourselves, is bad originality really preferable to a brilliant copy?

There are arguably a number of thematics associated with the pedagogies of re-use which we would like to outline. If anything, can this endnote position a few strategic ideas through which the fundamental principles of re-use pedagogic endeavours lie?

Anxieties

Who is appropriating whom?

Teaching re-use immediately requires participants to lock into systems of ownership, source, issues of appropriation, and who exactly is appropriating whom with what and then how? Re-use always asks the question, who is being recognised in this journey of remaking someone or other groupings voices/works and how, if at all, is that to be recognised? The complexities of these questions mean that re-use projects go far deeper than a blank-slate, new-build approach. In re-use, research for a project will often begin with the answer. We call this a reverse-normative approach. This is because if the answer is already supplied, then the design process really entails the formulation of which questions to ask in order for actions that can then be applied. Whether an existing building, a tile extracted from a redundant interior or a pile of surplus textiles are the project, the answer, then, is what actions will be applied to them through the questions raised by the designer(s). So, in re-use, the existing material in front of you provides all of the answers to the project in the shape of what tools will you develop to interrogate it with.

Originality/authorship

Working with the work of others is a direct riposte to the traditional ways of thinking about educating designers, where value is often attached to 'originality', 'authorship' and the daunting prospects of unfettered innovation. The associated anxieties around doing something that has not been seen or done before is a false-fallacy. It is a condition promulgated by the traditions of educational institutions where fears of plagiarism still haunt the corridors and the simulation or the copy of something is still regarded as taboo. How many conversations have you had with your students with regards to them being worried about the similarities of their work to something they have seen online, usually, to be fair, found on Dezeen or Pinterest. I have tutored students with 'invention-syndrome', a non-medical paralysis which is best cured with the soothing words that there is actually nothing new, just iterations of a number of ideas, developed until they become something that you consider your own. It is not uncommon for 'the precedent' to be acknowledged as a referent for the hallowed idea and the required technical detail for a floor section or cladding can be extracted wholeheartedly and repeated. How do we relieve our students of their anxieties around authorship?

Demolition Extraction

In the 21st century, the demise of extractivist approaches with which to materialise our environment means that teaching now has to be focussed solely on the re-designation of all existing matter. New-build and single-use processes will be obsoleted distinctions for making cities, buildings, interiors and artefacts. *The Pedagogies of Re-use* takes the position that in a world without demolition or discarding, what is already extant will provide the material for the profound transformation of the existing into the new. The remit of the book is to explore how academics across the world are teaching these processes to the new generations of designers and architects.

Sensibilities

Most of the companies and individuals attempting to divert vast amounts of the so-called waste (we call it resource) from landfill or incineration are not designers. Most often re-use in the construction industry IS the responsibility of others, that is the companies and individuals interested in resource management. However, as you will see within the pages of this book, there is obviously a huge opportunity for designers to bring our skill sets to the world of re-use. To do this, we need to learn new 'ways of doing', to create unique sensibilities which are hard to develop. For example, we need to appreciate that when we work on an existing building, or even a green field site, we are part of a long lineage of interventions, via human or natural causes, going back decades if not centuries. Given the opportunity intervening with a situation today is a huge privilege, and surely one that needs to acknowledge the value of past interventions, whilst respecting the need for future interventions to our own proposals. Acknowledging this idea of a propositional lineage enables us to appreciate the past, and perhaps to anticipate some themes and challenges that may present future design teams.

This book represents only the first thoughts and considerations around 'what does a practice of re-use look like?' What is the material, aesthetic and spatial potentials of creative re-use and adaptation? Some of these issues are addressed within the pages of this book, but these are early days. We hope that you are inspired enough to engage with the biggest existential challenge of our time; the climate and ecological emergency, and to consider ways to develop the re-use pedagogies and practices required to meet this.

A note on activism

Whilst writing this reflective statement in September 2023, the UK's Prime Minister, Rishi Sunak, is making daily statements rolling back on climate-sensitive legislation. His strategy is to try and appeal to the working-class voters that voted for his party back in 2019. Laws demanding that landlords insulate the homes they rent have been reversed. Coal mines and oil drilling licenses have been offered to fossil fuel companies, whilst value-added tax (VAT) at 20% remains on all retrofit (sorry Graeme) projects whilst new build is zero rated (0%). And all of this in a year of unremitting climate induced forest fires, floods and storms. On the upside, this may be the last days of a neoliberal right wing government intent on shaming anyone wanting to do something about the climate and ecological emergency. In the UK, we find ourselves in a period in time where climate activists are assaulted, arrested and even imprisoned for interrupting the everyday lives of 'normal' working class people just trying to get on with their lives – the very people and communities who will feel the real pain of climate change first (heat, fire, flood, food scarcity, energy crisis, etc.). Today, as I (Duncan) write this, there are people in prison because they took non-violent action in the name of insulating our homes. What times we live in.

We would suggest that an aspiration to develop a healthy re-use pedagogy is not a done deal. Despite many people, high profile, successful and respected people even, supporting this new sensibility, there are many who just can't see beyond what they already know. Asking questions, pointing out the (to us) bloody obvious can feel like getting in the way of a well-oiled machine (no pun intended) designed to satisfy the time-focussed, short-term profit-making aspirations of their clients – clients that must satisfy shareholders – we all know that game. However, things are definitely changing for the better. More multi-national corporations realise that even they can't run away from the consequences of 250 years of taking, making and throwing away or the industrial revolution as it is better known. Perhaps climate denial, or to be more precise for current times, putting off the decisions required today to reduce the impact of the climate and ecological emergency, is a more parochial pursuit proffered mainly by politicians across all parties trying to convince us that we can get back to the good old days. Well, they were never so good.

To quote from Scott McAulay's chapter '*Build Lifeboats Not Coffins*' earlier on in this book:

> Students have not been prepared for their unpredictable futures, and design studios continue to be run in a vacuum – insulated from the cultural, economic, environmental, technological, and political realities of the world outside the university environment. . . . We will know that architectural education is moving in the right direction when students leave it feeling as if they have the capacity to radically improve the world around them, despite the house being on fire.

Note

1 www.dezeen.com/2017/07/18/phineas-harper-opinion-copying-originality-architecture-assemble-cineroleum/

INDEX

Note: Page numbers in italics indicate figures.